Casting Homeward

An Angler and Naturalist's Journey to America's Legendary Rivers

Steve Ramirez

Illustrations by Bob White
Foreword by Richard Louv

Essex, Connecticut

An imprint of Globe Pequot, the trade division of The Rowman & Littlefield Publishing Group, Inc.
4501 Forbes Blvd., Ste. 200
Lanham, MD 20706
www.rowman.com

Distributed by NATIONAL BOOK NETWORK

British Library Cataloguing in Publication Information available

Library of Congress Cataloging-in-Publication Data
Names: Ramirez, Steve, 1961– author.
Title: Casting homeward : an angler and naturalist's journey to America's legendary rivers / Steve Ramirez ; illustrations by Bob White ; foreword by Richard Louv. Description: Essex, Connecticut : Lyons Press, [2024] |
Summary: "Writer, naturalist, and educator Steve Ramirez takes the reader on a physical and philosophical journey to some of the most legendary rivers and wild landscapes in America. Imbued with fly fishing throughout, this journey will seek to explore what makes certain places feel magical and meaningful—they are iconic waters"— Provided by publisher.
Identifiers: LCCN 2024002169 (print) | LCCN 2024002170 (ebook) | ISBN 9781493077694 (hardback) | ISBN 9781493077700 (epub)
Subjects: LCSH: Fishing—United States. | Native fishes—Conservation—United States. Classification: LCC SH463 .R345 2024 (print) | LCC SH463 (ebook) | DDC 639.20973—dc23/eng/20240410
LC record available at https://lccn.loc.gov/2024002169
LC ebook record available at https://lccn.loc.gov/2024002170

∞™ The paper used in this publication meets the minimum requirements of American National Standard for Information Sciences—Permanence of Paper for Printed Library Materials, ANSI/NISO Z39.48-1992.

To our one true Home, this beautiful and bountiful blue planet that we call Earth. And to the Home of our creation, the ideal and reality of a healthy, thriving community of human and nonhuman living beings. We Travel Together.

In the moment, is it still possible to face the gathering darkness and say to the physical earth, and to all its creatures, including ourselves, fiercely and without embarrassment, I love you, and to embrace fearlessly the burning world?

—Barry Lopez

Contents

Foreword vii

Prologue xi

Chapter One: First Cast—Guadalupe River, Texas Hill Country . . 1

Part I: Bristol Bay, Alaska 13

Chapter Two: Life and Death on the Agulowak River, Alaska .15

Chapter Three: Upnuk Lake by Floatplane, Alaska. 29

Chapter Four: Floating on Alaska's Ungalithaluk River. 41

Chapter Five: Floating on Alaska's Good News River 53

Chapter Six: Alaska Epilogue: Return to the Agulowak 67

Part II: Montana's Paradise Valley and Beyond 75

Chapter Seven: Wading DePuy Spring Creek, Montana 77

Chapter Eight: Floating the Big Hole, Madison, and Missouri Rivers, Montana .91

Part III: Grand Teton and Yellowstone Country, Wyoming. . . . 105

Chapter Nine: Wading the Firehole River, Yellowstone National Park, Wyoming. 107

Chapter Ten: Walking and Wading Teton and Yellowstone Country 119

Chapter Eleven: Floating the Snake River, Grand Teton and Yellowstone Country. 129

Part IV: Classic Waters in the Commonwealth—Pennsylvania Spring Creeks . . . 139
Chapter Twelve: Wading the Upper Little Juniata . . . 141
Chapter Thirteen: Wading Upper Penns Creek . . . 155
Chapter Fourteen: Floating Lower Penns Creek. . . . 167
Part V: Birthplace of American Fly Fishing—The Catskills and Northeastern Forest. . . . 179
Chapter Fifteen: Floating the Delaware, New York . . . 181
Chapter Sixteen: Wading the Esopus, New York. . . . 193
Chapter Seventeen: Wading the Farmington, Connecticut . . . 205
Chapter Eighteen: Wading the Battenkill, Vermont . . . 217
Part VI: Casting Homeward—The Foothills of the Ozarks and the Heart of Texas . . . 227
Chapter Nineteen: Smallmouth Bass Fishing, Eastern Oklahoma . . . 229
Chapter Twenty: Floating the Llano River, Texas Hill Country. . . . 245
Chapter Twenty-One: Just One More Cast—Devils River, Texas Hill Country. . . . 253
Epilogue . . . 265
Acknowledgments . . . 267

Foreword

Old Norman Maclean was on to something important when he wrote, "In our family, there was no clear line between religion and fly fishing." Maclean's father, a Presbyterian minister and fly fisherman, told him and his brother that Christ's disciples were fishermen, and the two boys "were left to assume . . . that all first-class fishermen on the Sea of Galilee were fly fishermen and that John, the favorite, was a dry-fly fisherman." An oblique spirituality runs through Maclean's famous book *A River Runs Through It*.

This is the same river of spirit that runs through the pages of *Casting Homeward* and the prior books in this series: *Casting Forward: Fishing Tales from the Texas Hill Country*, *Casting Onward: Fishing Adventures in Search of America*, and *Casting Seaward: Fishing Adventures in Search of America's Saltwater Gamefish*. These books place Ramirez in the elite ranks of a group of writers who focused much of their work on angling, including Maclean, Dave Whitlock, Ted Williams, John Gierach, Nick Lyons, and Ernest Hemingway. These were authors who never wrote only about hooks and lines and the fish that got away, but also about the deeper currents of human experience within the spirit of nature. Like Maclean's father, Ramirez is mostly a dry-fly fisher, though not against using a dropper and deep-running nymph, particularly when anglers within sight are catching and he's still fishing.

As a man and a writer, Ramirez favors questions over answers, and as Bob Dylan wrote about a former girlfriend, "knows too much to argue or to judge." He is, in fact, ambivalent about the subtleties of fishing. He is sympathetic to and respectful of the fish he catches and gently releases, but not against occasionally frying a trout streamside. Either way, he always gives thanks. *Casting Homeward* invites and guides the reader on an epic journey to some of America's most evocative and fragile and legendary rivers. The Guadalupe, Ungalithaluk, Good News, Big Hole, Madison, and Snake Rivers—and many more—and then comes home to the Llano and the Devils River in his beloved Texas Hill Country. While doing so, he travels backward in time, retracing the westward trek of Euro-Americans and exploring the wisdom of Indigenous people whose ancestors were the first to fish in North America.

Along the way, he does describe the intricacies of fly fishing, but also the longing and pain that so many anglers cast upon the waters, sometimes successfully. In addition to being a poet and author, Ramirez is a former Marine and law enforcement officer. For years, he has suffered from post-traumatic stress disorder. At the outset of his travels for this book, he learned that he has heart disease. Years ago, another fine writer and fly fisher, Margot Page, shared this thought with me: "I almost hate to call it fishing. I'd rather call it water treatment. Yes, it's about the line and these wild flashes of light you see in the stream, but it's really the water that we go to and the water we've always gone to." He would agree with Margot Page's prescription for body and spirit. (Many of Ramirez's friends on the water are women, a reminder that for years women have comprised the fastest-growing demographic in fishing.)

He writes, "Whenever I walk into a river, I do so with reverence. Nature is my sacred place. Nature is my home. I imagine God as an outdoor enthusiast—like me. If anything was made 'in his image' it must be rivers. They too have no real beginning or end. They live on the land in several seasonal forms, and in the sky as a mostly invisible member of the holy trilogy of water, vapor, and ice."

Such reverence fuels not only a writer like Ramirez, but also many of the conservationists who fight to release dammed streams and protect the waters, and to deal with the even larger challenges of biodiversity collapse, climate change, and the kind of development that beheads mountains and grades forests. These efforts are not only about defending and replenishing the natural physical world, but about the spirit that runs through it. "Just because something is invisible doesn't mean it's not there. I can see how the air moves the clouds, trees, and grasses. I can feel how gravity holds me close to the earth—rooted without roots," he writes. "Every heartbeat on the earth is my heartbeat as well." You feel that heartbeat in his words, and you feel the arc of the line as it floats over the waters, and the dry fly above as it skitters on the waves, and the nymph when pulled down and away by those deeper currents.

As Ramirez finishes his most recent journey, he returns to the Hill Country, where he releases into a loved river the ashes of another author and angler. Ramirez writes movingly of home and away, always with empathy and gratitude and humility. "I am, after all, an imperfect but persistent Texan Buddha," he writes, adding, "If you are reading this, it means that you have chosen to cast your line beside me."

—Richard Louv, author of *Our Wild Calling* and *Last Child in the Woods*

Prologue

What happens when people open their hearts? They get better.
—Haruki Murakami

This is a story about the power of perspective, purpose, and choice. It is a journey through what is and into what could be if only we changed our perspective, refocused our purpose, and made new wiser choices—together. And this is a sojourn into a realm of discovery where we seek to live healthier, happier, more meaningful lives—that are as connected to Nature as Nature is connected to us. This is a story of enlightenment, empowerment, and empathy. All of this . . . in a fishing adventure.

Chapter One

First Cast—Guadalupe River, Texas Hill Country

Every step taken in mindfulness brings us one step closer to healing ourselves and the planet.

—Thich Nhat Hanh

When I was diagnosed with heart disease, I went fishing. I went fishing, not because in doing so the blockage in my coronary artery might magically vanish with each new cast, or the birth defect that the cardiologist discovered might somehow grow stronger and reroute itself like a river that changes its channel after a storm. I knew these things cannot come from a tight loop or a good drift. Instead, I went fishing because I knew that the river has the power to heal me in deeper ways, and that true friendship shared on the river acts to remind us that we travel together. We are not alone in our aloneness.

As I loaded my truck with fly-fishing gear and provisions on a brisk and bracing morning, I felt grateful for every life-sustaining breath. We're all in this together. I exchange elements of existence with the trees. They give me oxygen and I give them carbon dioxide. Together we thrive, each indebted to the other. Reciprocity is the root of natural law—take only what is freely given and give what is truly needed.

My mug of rich, black, Peruvian coffee tasted so intense that it seemed somehow more glorious than normal. There's nothing like being reminded of the singular specter of death to focus us on the many blessings of life. Awareness of impermanence is a gift, if we don't dwell on it but, rather, learn to live with it.

I drove eastward toward the golden sliver of the sunrise's first light, which appeared to set fire to the naked limbs of live oaks and meadows of undulating prairie grass. The Texas hills were bathed in the morning glow, even as behind me they still contained the last flickers of distant stars. I knew that the same starlight would continue to reach the earth's daytime skies well after the stars that sent it have become extinct. Light that travels through darkness tends to outlive its origins. This is true within the cosmos and within the pages of a book. But the starlight will become invisible to my human eyes and mind in the brilliance of our own single star. Invisible—like air, gravity, love, and whatever it is that we sometimes call "God."

Just because something is invisible doesn't mean it's not there. I can see how the air moves the clouds, trees, and grasses. I can feel how gravity holds me close to the earth—rooted without roots. I can notice how love is reflected in the eyes and actions of another living being. And I can see in Nature the creations of something far more vast than mere human imagination can conceive. We all struggle to hold on to what cannot be held. Still, the invisible carries songbirds up into the sky and raindrops down to the river, and me onward toward feelings of hope, joy, and gratitude. It was a beautiful morning, and I was once again simply happy to be alive.

I have an ambiguous relationship with the word "friend." Like the word "love," people seem to toss the word "friend" around without thought or considered meaning. I don't. For me a friend is someone I trust, value, respect, and care for deeply. For me, love is unconditional and eternal. I never understand how people can turn friendship or love on and off like a spigot. For me, like a river, it flows—even when sometimes it is seemingly invisible. Always below the bedrock of my soul, there is love.

When I arrived along the banks of the Guadalupe River and pulled in front of the L&L Campground, I immediately saw that Cari Ray was already waiting for me in her dark green pickup truck with pop-up camper with its door decal that reads, "Fisher of Zen." Cari is my friend.

She is a peaceful soul who, like me, has fought through hardship and heartache while somehow remaining intact, entire, and empathic toward other souls—human and otherwise. Whenever we see each other, we smile, embrace, and invest a good bit of time tailgate-talking about life in general—catching up before we begin catching fish. It just feels right.

When we arrived at the river, it was running cold and clear and glistening in the early morning sunlight. It had the look and sound of broken bits of sea glass that tumbled with the breeze. Just a week prior, the ancient cypress trees that line the river were splashed in yellow-orange autumn colors. Now those leaves that once captured the sun to create edible energy were drifting with the currents of water and sky. I looked down at my feet, submerged in the world of trout and nymphs, and then up above my head into the world of warblers and mayflies. Whenever I walk into a river, I do so with reverence. Nature is my sacred place. Nature is my Home.

I imagine God as an outdoor enthusiast—like me. If anything was made "in his image" it must be rivers. They too have no real beginning or end. They live on the land in several seasonal forms, and in the sky as a mostly invisible member of the holy trilogy of water, vapor, and ice. They too are eternal, ever changing yet unchanged. Nature prays through the sounds of falling raindrops, rushing rivers, and rolling seas. I pray that way too. Water, wind, and birdsong connect me to the divine.

We walked downcurrent toward a nice run that Cari thought might contain a few fish. It wasn't enough water for both of us to fish, so I invited her to take the first cast of the morning. We were using Cari's hand-tied rigs with a midge nymph suspended below an egg fly and a nontoxic split shot. She ties them so that you get an even drift, and we used indicators that were easily adjusted for water depth as we moved from one run to another. Cari is a better angler on her worst day than I will ever be on my best. But most importantly to me—she "gets it." Neither of us count fish caught or care a hoot about who catches the biggest fish. We simply enjoy being alive and outdoors.

It took only a few presentations for Cari to hook and land a couple of smallish trout, but she also missed a strike of another that seemed to own that slice of river. That's when she decided to play a game that we both enjoy. We call it, "I'm going to catch *that* trout."

That trout doesn't necessarily need to be the biggest to be chosen as the challenge at hand. It simply needs to feel somehow special. It needs to show a bit of unique character or color or even innate wisdom as it rises and then rejects our offering. I guess we try to catch and then release it because it is somehow "love at first sight." It just feels like it's meant to be. It's as if a young man standing in a subway car gets a glimpse of a young woman in the next car just as she makes eye contact, smiles, and looks away. And all at once he is committed to unlocking the puzzle that separates him from the girl with the fleeting smile. He must meet her, even briefly. He waits for the train to pause and the doors to open—and then he casts his line . . . hopefully. This is how it feels with *that* trout. It feels like destiny.

So I stood in the river just watching my friend as she worked the quick water again and again, adjusting to the occasional flash of the fish and the frequent visual silence of the bit of pocketwater he called home. Then, all at once, it happened. The fish accepted the offering, the indicator moved like the opening of subway doors, and my friend lifted the rod tip just in time to catch and release *that* fish. And when I tell you it was beautiful, I mean both the big bright rainbow trout and the moment when he swam back home, seemingly no worse for the meeting. Cari and I smiled at our good fortune as we waded downriver together toward the pool we were hoping to fish. It was a pleasing beginning to a beautiful day.

When we arrived at the pool that resides at a bend in the river, we found its upper stretch occupied by a man named Malcolm. Cari asked, "Any luck?" Malcolm smiled and said, "I hooked and lost two fish, but none have been landed yet." We asked if he minded us working the downcurrent section and he said, "No problem. Maybe you'll have more luck than I did." Malcolm was a nice guy. I instantly liked him and found myself wishing he'd land a fish.

The bottom section of the pool had quite a few fish stacked up, most likely picking up midge nymphs as they flashed and fluttered in the current. There was only enough room for one angler, so Cari and I decided to take turns—one fishing as the other watched. I love fishing with friends who are generous of heart, and yet secure enough to accept and recognize the value of mutual generosity. If only more of the human world saw life as abundant rather than scarce and accepted kindness as the first rule of living. How cool would that be?

It only took three passes before Cari hooked and netted a nice fish. Then it was my turn, and I caught one too. Then Cari caught another, which made several acrobatic leaps before coming to her net. We continued like this for quite a while, until it became obvious that this section of the river was done for the moment and the trout needed a break. All this time I was silently hopeful for our new friend Malcolm, but he had caught nothing even though we could see the fish stacked up in his section of the river. That's when he said, "Y'all are making me feel bad! I've been here for a couple hours and caught nothing; you came along and seem to catch a fish on every other cast. Do you have any pointers for me?"

Cari is not only a wonderful angler, professional fly-fishing guide in Texas, New Mexico, and Colorado, singer-songwriter, and friend, but she is also a kind soul. With Malcolm's permission she looked at his rigging and made some suggestions on both setup and fly selection, and then gave him some quick instructions on presentation. That's about the time that Lew came wading downriver and walked up to me for a quick chat.

I'd noticed Lew earlier that morning. He wasn't fishing when I saw him; he was walking upriver as if going on a morning stroll down a neighborhood street. And he walked up to us in quite the same way, as if we were meeting on some city sidewalk while having a morning stretch of the legs. With his first few words it was apparent that Lew was a native of New Jersey who had chosen to retire in the Lone Star State. He and Malcolm were fishing buddies who had driven over from Houston, where they both now live. Like Malcolm, Lew was a nice guy who was quite involved in Trout Unlimited and concerned about the changes they'd witnessed on their favorite rivers and in the nature of Nature in these modern times. We chatted for a while before saying our farewells, giving both Malcolm and Lew a bit of space to figure out the mysteries of the river-bend pool before making the long drive back home.

I've met some of the best people while standing in a river. The rare few jerks I encounter are more tragically comical than anything. No one is as empty as someone who is full of themselves, and I always find myself walking away from those folks wishing that they could realize the reflection off the water is not the river. They are so often standing in the "water" that they should be casting toward. We're all just trying to catch up with our own true selves. As Ram Dass once wrote, "We are all just walking each other home."

Cari Ray waded upriver toward another long pool at the end of a wide, well-aerated riffle, and I decided to snip off my nymphing rig, change my leader, and tie on an olive micro-streamer that was tied by my friend Aileen Lane. We had used these to great effect while chasing Gila trout in Arizona and redband trout in Idaho, so I wanted to give them a try here.

I began working a few deeper areas of fast water so that the fish wouldn't get a long look at the streamer and might make a quick impulse grab, and it worked, with me getting and missing two half-hearted strikes, and ultimately hooking and losing a big fish that I foolishly tried to horse into the slower current—while breaking my 5X tippet in the process. I always feel bad if a fish gets free with my fly still in its lips. It makes me ask myself once again, "Is it right for me to be hooking fish for my own amusement?" I do think about this from time to time, and I thought of it for a little while as I tied on a white Woolly Bugger and waded over to the next fast-water run.

At first, I just watched the water and searched for any signs of fish beneath the foam line. I envisioned myself as the fish, choosing the best spot to hold and hover as the river brought me oxygen and food, and then I became an angler again and cast my line so that my streamer would swing to the place where my imaginary fish was swimming. And that's when he left my free-flowing imagination and entered my current reality. The strike was swift and solid, and my hookset came naturally as if our meeting was predestined. He was a big, muscled fish with bright sunlit colors and a proclivity for alternating between powerful runs and breath-taking leaps. When I finally had him in hand, I freed him as quickly as I could. He was magnificent, and I was mesmerized by the sight of him sliding back into his place in the river—a deep place among the rapids where time and timelessness meet. As always, just before setting him free, I whispered "thank you." Gratitude creates joy, and I am a joyful man.

In her wonderful book titled *Braiding Sweetgrass: Indigenous Wisdom, Scientific Knowledge, and the Teaching of Plants*, author Robin Wall Kimmerer speaks of "the grammar of animacy." She writes of her effort to be "bilingual between the lexicon of science and the grammar of animacy." And she points out that in the languages of English or science, "you are either a human or a thing." Finally, she asks the question, "Where are the words for the simple existence of another living being?" Like me, she sees the world as a community of human and nonhuman living beings. We see the souls that reside all around us.

After taking a few casts toward a different and shallower pool, without result, I decided to go back to the one where I had just caught the chunky rainbow and give it another go a little farther downstream. On about the third cast I saw a flash beneath the fly but no take, so I cast again and then again until the flash became a take, hookset, and landing of another pretty fish. He wasn't quite as large as the first but was every bit as beautiful, and I could feel myself smiling as he swam back to the place in the river where we first met. I wished him all the best, even though I knew his ultimate destination. Someday he will rejoin the river in a different form—just like me. I suspect neither of us is in any rush for that day.

I guess I've always known that I'm different from so many of the good people that I am fortunate enough to meet while standing in a river, or walking down a canyon trail, or sitting beside a sunny window in a neighborhood cafe. My brain is wired differently, not only because of my nearly lifelong journey with post-traumatic stress disorder (PTSD), but also because I think in poetry and metaphor. I see the world in spiritual animistic ways, where everything is alive and soulful.

Since childhood I have felt myself flying with the songbirds and swimming with the rising trout. I remember as a boy shooting and killing my first dove. I held its heart in my boyhood hand and when I looked at it, I saw my own—still beating in my chest. Whereas so much of humanity seems to avoid the thought of mortality and of being a part of Nature rather than apart from Nature, I have always felt the entire mortal journey playing out at once, with cognition of the connection between the being that is me and the beings that are currently living in another form.

The earth breathes through the leaves of trees and the plankton-filled surface of the sea. In an alchemistic partnership with plants, water, and soil, it transforms our star's light into our planet's life. And in doing so, every heartbeat on the earth is my heartbeat as well.

I looked upriver to see my friend Cari casting and catching one fish after another. She was just above the fast water, tending to a long pool where the trout were stacked up, hungry, and not selective. These trout were all in the sixteen- to twenty-one-inch range, which the local Trout Unlimited Chapter had stocked here, along with the ten- to twelve-inch fish that Texas Parks and Wildlife stocked by the truckload. The U.S. Army Corps of Engineers dam just upriver has created a tailwater fishery that is as artificial as all tailwaters are, and these fish were anything but wild—quite like most humans. Like most of us, they had not yet discovered their true self. I felt a bit sorry for them—and us. But we all have our chance to go wild if our heart still beats and the light of the universe still shines in our eyes.

Wading downriver, I found a bit of quick water with some deep pools and cutbanks and began casting the white Woolly Bugger down and across hoping to elicit a strike. It was a narrow place, lined with tall and ancient cypress trees dripping in sunlit Spanish moss. It looked like living tinsel. Few people ever notice the tiny azure flowers that these unique air plants produce each spring. Every living thing wants to keep living in one way or another. Sometimes I think DNA carries more than genetic codes—it carries memories.

Just downriver from me was another angler. He was standing alone along a line of anglers, not lonely but in peaceful solitude. There were several anglers just upstream of Cari, and we each took turns greeting each other and allowing the other to experience the oneness with the river that only fly anglers and kayakers might know. We adjust to its moods and listen to the stories it can tell and the lessons it can teach. We look deep beneath the surface as best as we can to learn and understand its many attributes. We learn to recognize the meaning behind riffles, runs, plunge pools, and pocketwater. We lift stones to see who else lives there—salmonfly, caddis fly, mayfly, dragonfly, damselfly, crawfish, sculpin, or minnow. And through it all, we do as we must with anyone that we truly love—we pay attention. We listen for the sound of its beating heart. We watch for the light in its eyes.

Most places I fish in the Texas Hill Country are remote, and much to my satisfaction, I am often the only angler on the river. I love fishing alone or with a special someone whom I invite to fish with me because they add to my experience—not detract from it. Cari is such a soul—she is a true seeker of Zen. And while I prefer the solitary streams and rivers, there is a sense of community here that holds its own charms. It gives me a flicker of hope for humanity and the earth that people of kindness and mutual respect still exist. And all this caused me to consider the journey ahead of me . . . the one you are about to travel with me. It will be imperfect—like me. But it will be what it will be, and I will do my best to practice acceptance, empathy, and gratitude along the way. I am, after all, an imperfect but persistent Texan Buddha.

If you are reading this, it means that you have chosen to cast your line beside me. Together we will experience waters that, for one reason or another, Euro-American humanity has labeled "legendary." And because they have been designated as such, they will often not be places where I can experience the solitary immersion in Nature that I so often crave. I suspect a Native American might feel more as I feel: that the river that is special to any of us is the one that we feel most a part of and have come to know intimately—the one we love. Fishing to catch big fish or many fish is akin to anonymous sex; casting for connection is akin to intimacy. In Nature and in human nature, I seek intimacy not anonymity.

Still, I am curious about what I will discover about the planet, humanity, myself, and life itself as I travel from the remote far western waters of Alaska to the historical streams of the Catskills that flow just beyond the skyline of Manhattan. What will I learn about America and Americans as I travel from those watersheds least impacted by humanity toward those most impacted—so far?

The fast water and its deep pools and ample cutbanks gave me not a single trout, but I didn't mind. I had fun practicing my steeple cast against the backdrop of cypress trees and bunchgrasses. And I enjoyed watching the angler just downstream who was gleefully unaware of my existence as he hopefully sent one soft cast after another into the river. So I snipped off my fly and reeled in my line and just stood there—somehow transported so that I felt as if I were the only human on earth and the river was my home. I looked down at the river wrapping around me, embracing me, and washing away my cares. I was right to come here, knowing that the river has the power to heal me in deeper ways and that true friendship shared on the river acts to remind us that we do indeed travel together. All too often, we fear what we need most—oneness.

After wading to the river's edge, I sat on a rock and simply watched the water flow by me and around my half-submerged limbs. They looked like the roots of the cypress trees, and in thinking this, I felt comfort. Cari began to work her way downriver toward me, fished a bit more in a quick water bubble line just across from me, and, after a while, reeled in and waded over to join me on the rock-strewn shore. We sat there in silence at first, eating nuts and raisins and allowing the peace of the moment to sink in. And then we spoke of my struggles with asthma, sleep apnea, PTSD, and now heart disease. And we spoke of the debilitating impact that long COVID has inflicted on her life and health. We shared our fears that we may no longer be physically capable of doing the things we once did and, by extension, living the lives we've always lived. But in the end, we spoke of the day we'd enjoyed together on the river, and of our friendship, and how fortunate we felt—simply to be alive.

It has occurred to me that just as our bodies and souls are living and breathing and beating in their own ways and rhythms, so does the earth and every living part of it. Every drop of water and every handful of soil is alive—or should be. In her book *Braiding Sweetgrass*, Robin Wall Kimmerer points out that in the language of her Native American ancestors, rivers are living beings, "imbued with spirit," not things to be used and used up. In her ancestral Potawatomi language, a person might consider the water of a river and say, "It is being a river." That same water could otherwise become a stream, bay, ocean, rain cloud, or glacier, but in this moment it "is being a river." That's how I see myself. In this moment I am being "human."

There is a great deal about living that rivers can teach us. They teach us how to bend and flow over, under, and around obstacles. Rivers know that the form and function they embody today might not be the same tomorrow. Rivers understand that change is as inevitable as it is magical. They adapt and re-create themselves in such a manner that no matter the circumstances encountered, they remain true to their original nature—they continue to flow. And these are good lessons for Cari and me—and you. We have nothing to fear.

As if completing a circle, I soon found myself driving westward toward a golden sliver of sunlight that once again seemed to set fire to the naked limbs of live oaks and intermittent meadows of bending prairie grasses. The Texas hills were bathed in late evening light, even as behind me the first stars appeared faintly, almost as if they were uncertain if this was their time to shine. It was. And that's the lesson I most need to remember in the coming year and years. The time to shine is now, and the place is here.

This is the wisdom of the river. We do not count the miles or the years. We do not worry about flood, drought, or how many breaths or heartbeats we may be granted. We simply carry on—being a river. In the end, there is no end. Raindrops falling need not fear the sea.

Part I

Bristol Bay, Alaska

Chapter Two

Life and Death on the Agulowak River, Alaska

The goal of life is to make your heartbeat match the beat of the universe, to match your nature with Nature.

—Joseph Campbell

It's 5:40 a.m., and I woke up with Alaska in my dreams. Alaska is like that. It hangs in my waking thoughts and calls to me in my sleep. I remember every detail. Like how the alders and spruce embraced the edges of its cold clean waters and how the salmon, char, trout, and grayling vanished into its deepest holes. I remember the sensation of floating on currents of air as the 1957 de Havilland Beaver carried me over forests and tundra, and then floating on the currents of its rivers while casting and connecting to the many beating hearts that swam within. I recall the primal gaze of brown bears and how that moment of eye contact grounded me in the certain knowledge that in the wilderness, I'm simply another source of protein. And at this moment in the darkness of my room, I am reliving my first and final days when a single river and special fish were both joined to the deepest part of my eternal soul. This is our story.

Ever since I was a boy who sat up at night reading Jack London, Russ Annabel, and vintage outdoor magazines, I've dreamed of this moment. To my childhood mind, taking off and landing in a floatplane was the ultimate gateway to adventure. It brought to mind the images of floating over rugged mountains, raging rivers, and endless expanses of wilderness where trees were many and landing sites were few and far between. And now here I was, about to begin my weeklong adventure at Bristol Bay Lodge with my buddy, sporting artist, author, and guide Bob White.

Ron Salmon is a quiet, kindhearted, and exceedingly competent pilot with over fifty years of experience. Every lesson he ever learned during his winged lifetime was self-evident as he effortlessly launched and landed the floatplane on the surface of Lake Ageknagik. The subtle separation between the surface of the lake and the underbelly of the plane's pontoons was surprising, and it took me a while to realize we were airborne as perspectives altered and realities shifted. As we rounded Jackknife Mountain, Ron adjusted flaps, mixture, propeller, and throttle so that the alchemy of his actions resulted in the pontoons resting back onto the water with the natural grace of a mayfly. Stepping onto the dock and into my childhood dreams come true, I could never know that at that moment, I was coming home.

The flat metal tin can of a johnboat bounded and slid across the choppy surface of the Agulowak River as my newest friend, Sam Fisher, deftly maneuvered us toward Grayling Island. I had barely stepped off the floatplane and into the boat when I told Sam that my number one dream was connecting with an Alaskan grayling. I asked if he thought we might find one before dinner, to which he responded, "Heck yeah!" Sam is a kind, open-minded, and able young man with an infectious smile and good nature. We hit it off immediately, and it mattered not if we were twenty-six or sixty-two—real friendship is timeless. It was my first day, and every possible reality lay in front of me as each cast would roll out like dice tumbling across a table.

Sam set the anchor at the edge of a community of half-submerged willows, and I began casting a fluffy white dry fly with a tiny Copper Bob nymph dropper. Almost immediately the dry fly became an indicator as I raised my rod tip a bit too softly, hooked, and then lost a grayling while we both lit up with excitement and then moaned with good-humored disappointment. That scene continued through four more half-hearted hooksets before I finally shook off the first-day jitters and landed the first grayling of my current lifetime. It was all I had hoped for, and more. He was a beautiful fish, and I tried to burn the image of him swimming back home into the currents of the Agulowak, where he would live and die and live again as his DNA passed through time. It is the way of things that we are all both mortal and immortal—all at once.

For me, a fish is not a thing; it is a living being with a desire to survive as long as possible and with the best quality of life possible—just like me. When I do take the rare photo with a fish, I am never intending to say, "Look what I caught," but rather, "Look who I met." This is why I bring "him" or "her" to the net—not it. And this is why I say "thank you" when I release each beautiful creature back into the river of life. In the language of Robin Wall Kimmerer's Native American ancestors, I might say, "megwech." The intention is the same. Respect, empathy, and gratitude.

For me, a river is not a thing; it is a living being with a desire to flow as long as possible with clarity and good health—just like me. I've noticed that how we treat ourselves often determines how we treat others and the planet. If I eat more plants and less meat and processed foods, my heart and soul will thrive. If I pollute my body with fat, sugar, and salt, I will no longer be the man I might have been. It's the same with forests and fish and rivers and relationships. We get what we give.

The morning came with coffee and breakfast and a view of Jackknife Mountain while looking across Lake Ageknagik from the window of Bristol Bay Lodge. I sat there in the soft silence with my friend Steve Laurant, who is the owner and general manager of the lodge, and Bob White, who is more like a brother to me than simply a dear friend. We didn't need to say much beyond our usual morning greetings. We sat together in silence or spoke in hushed tones about the peace of the moment and the joy of being alive. We didn't even talk about fishing. Among friends, silences are at least as important as spoken words. There is a comfort that comes with sipping coffee in the morning beside a person or persons who see the same magnificence and magic in the world around them and within the universe.

The Agulowak River is the home waters of Bristol Bay Lodge, as it is just across the lake from the dock that harbors several boats and three de Havilland Beaver floatplanes. The three full-time pilots include Steve, Ron Salmon, and a man who goes by the call sign of "T-Bird." (More on T-Bird in another chapter.) The Agulowak is approximately four miles long and is the connecting river between Lower Nerka Lake and Lake Ageknagik as part of the Wood River watershed. It is renowned for its large native rainbow trout, Arctic grayling, Arctic char, Dolly Varden char, and as the spawning grounds for approximately 200,000 sockeye salmon, with another 2 million passing through its waters on their way farther up the drainage.

After breakfast, we looked at the chalkboard on the wall to see what the day might hold, and next to my name was the term, "Wok." This meant that I'd be fishing the Agulowak, so I'd be loading onto a boat and not a floatplane. Bob and I walked down to the docks and met our guide for the day, a nice young man named Ethan Warren. Ethan was calm, quiet, patient, and competent with an easy smile. I liked him. The boats were basic sixteen-foot metal johnboats with oars on the sides and enough motorized horsepower on the back to get the job done without being obnoxious about it.

If taking off in a de Havilland Beaver is my favorite way to launch into adventure, then crossing open water in a boat is a close second. There is something magical about the bouncing of the bow across the water, the coolness of the air and spray on your face, and the anticipation of what lies ahead. And there is something meaningful about looking back at the wake behind the boat and the changing perspective of the place you left behind and the one you are growing ever closer to discovering. It's not that you wanted to go; it's that you felt compelled to go. As if staying behind would cause your moist lungs to stop breathing and your warm heart to cease beating. Living life urgently demands that we live gratefully in the moment while looking forward and remembering the lessons of our yesterdays.

Our target species on this first full morning was sockeye salmon. and the location was a turn in the river known as the "sockeye tree." The salmon are everywhere in the Agulowak as they swim almost zombie-like against the current toward their destiny—the place where they were born and where they will spawn, die, and give their last full measure by adding nutrients to the river and surrounding forests. Salmon and coastal brown bears are integral parts of this ecosystem. Without them, it is forever diminished. These fish are the stuff that such dreams are made of. They are the brick and mortar that built this wilderness. They are its lifeblood.

As we pulled up to the sockeye tree, we could see that my new friends Pue Nguyen and Steve Negaard had already arrived and were busy casting and catching as waves of salmon swam over the shallow water that surrounds this bend in the river. And it was this shallow water that made this act of catching these salmon possible. The rigging we were using was basically a salmon egg bead suspended above a hook with a "Slinky"-style weight attached to a swivel clip. The weight allows your rig to bounce along the bottom; once the angler feels the pull of the fish, the hook is set and hopefully the fish is landed. While the sockeye salmon are predominantly plankton feeders and therefore less likely to strike, the bead allows the angler to simultaneously fish for the Arctic and Dolly Varden char and rainbow trout that may be following the salmon in hopes of eating eggs that may be washed downcurrent from the redds. Beyond their platonic feeding habits, sockeye salmon are unique in that they require spawning rivers and streams that have lakes at their headwaters—such as the Agulowak, which empties into Lake Ageknagik from its headwaters at Nerka (the Russian name for the sockeye salmon) Lake.

This being the first full day of fishing in a new place, I was focused on getting to the boat and out to the river, which led to me forgetting to bring a few items that later proved meaningful. They included gloves, a head net, and mosquito repellent. I never made that mistake again. I quickly concluded that the most frightening sound in the Alaskan wilderness is not the growl of a bear or snarl of a wolf—it's the whine of a tiny bloodsucking insect in your ear. Enough said.

Upon arrival at the sockeye tree, the guides set up a plastic folding table and laid out a fillet knife and a club for dispatching the fish we would be keeping today. The fillet knife was long and thin and sharp—like any other of its kind. The club was part of a moose's femur bone—a truly Alaskan choice. Although I'm usually a catch-and-release angler and go to great effort to be careful with the fish I catch, I am not opposed to a fresh salmon shore lunch when the run of fish is healthy and plentiful. If bears knew how to fry potatoes, I'm sure they would. We are all hunter-gatherers by nature.

We lined up along the bank with me being the farthest downstream, Phu Nguyen just above me on the other side of the tree, and Steve Negaard just upcurrent of Phu. Bob chose not to fish and instead gave me tips for the best way to cast and land these fish, as this was all new to me. Bob has been coming up here for a couple decades and guided on this and other rivers for over a decade, so I was once again the benefactor of his seemingly boundless generosity. I am forever grateful.

The sockeyes seem to arrive in surges and waves of fish with relatively quiet moments in between. The trick is to cast across and slightly upcurrent as the salmon swim by and then follow the arc of the line as it swings down and inward toward the bank. You do this until you feel the tug of a fish and set the hook with some authority. My first fish of the day was a beautiful Arctic char that took the salmon egg imitation. I landed and released that fish quickly, and soon after was into my first bright sockeye salmon, which after a brief battle was inside Ethan's net and quickly processed into fillets in the cooler. Sometimes the circle of life ends in a frying pan over a campfire, and there's nothing wrong with that in my world when done ethically.

I began getting the hang of this technique and was catching sockeyes with some degree of regularity, as was my friend Steve Negaard. But our friend Phu was in some sort of sweet spot and his mojo was working because he quickly gained the name "one-cast Phu"—it seemed as if he was catching a fish on every cast. It didn't take long for all of us to limit out on sockeye salmon. The fishing was methodical and the catching was frequent, so in short order we had to adjust tactics and target different fish—rainbows and char.

To target the rainbow trout and char that were themselves targeting salmon eggs and the other potential gifts from the spawning process, we switched to a nymphing system that included a "flesh fly" and an "egg fly." The egg imitation is most often a small plastic bead set a few inches above the hook on the tippet of the leader. Some egg flies are made of yarn, but the principle is the same: You're hoping to induce a strike from one of the predatory fish that are feeding on stray salmon eggs. The flesh fly imitates the bits of salmon flesh that float off the bodies of post-spawning fish while they are in the process of dying or after death. As noble as a rainbow trout seems, it's not above eating carrion. The process is much the same as for salmon—casting up and across the current and watching the indicator as it drifts freely downcurrent. The strike is a steady submersion of the indicator that gives the angler a second or two to set the hook. Sometimes I took four seconds and set the hook only to find emptiness at the end of my line—just like in life. Half of success is that your ship comes in and the other half is that your bags are packed. Or as my dad used to say, "He who hesitates is lost."

About the time we were wrapping things up at the sockeye tree, a boat came by and the guide piloting it called out to us that a large brown bear sow with two cubs was about two hundred yards downriver from us and heading in our direction. I had noticed when we arrived that the black mud along the shoreline was covered in bear tracks, and although I was keen to see my first Alaskan coastal brown bear, I was cognizant that up close and personal with a momma bear and her cubs wasn't the ideal introduction. Based on the calm urgency of the guides as they cleaned up the area of fish guts and began our movement offshore, they concurred with my conclusion. We never did see that brown bear and her cubs that morning. I suspect she did see us.

The day prior while Sam and I were chasing grayling, a few of my fellow anglers were at the sockeye tree, and they reported witnessing an epic battle between a large sow with cubs and a big boar that was trying to get at and kill her cubs. Momma bear won the day. The male bears will kill the cubs of another male in an effort to force the female back into "heat" so he can have the opportunity to mate with her and pass on his genetics into the future. I can only surmise that a momma bear has either a short memory or a deeply forgiving nature if she is willing to mate with the male who killed her offspring. Then again, I'm not a bear. Who am I to judge?

As a man who aspires to imperfect Buddhahood and who invests a great deal of time and energy in reflection and resurrection, it's easy for me to forget that Nature is a complexly simple system of living beings all trying to survive and pass on their genetics through time—and that's it. I'm not sure how sentimental a warbler is about caterpillars or how well a contemplative wolf might do in its dog-eat-dog world. I doubt a grayling ever feels regret for swallowing a mayfly, and I suspect that a bear barely thinks about the hopes and dreams of the moose it ambushes, kills, and eats. But me, as a philosophical primate with opposable thumbs and a penchant for seeking something we call "wisdom," I'm stuck with an overabundance of existential angst and a lifetime of trying to solve for "why?" Is there more wisdom in living like a warbler? I wonder if the monkeys in a zoo ever find us humans to be amusing. "Human see, human do."

After catching my limit at the sockeye tree, Bob and I decided to focus on trout and grayling for the remainder of our first day. Ethan acted as a human anchor by standing in the river's shallows and walking us slowly downstream as far as he could, calling out likely places for fish on either side of the boat. Bob has guided on the "Wok" for years, so he didn't need the help, but I did.

My least favorite form of fly fishing is nymph fishing with an indicator, but that was exactly what we were doing because it's what the Agulowak and its fish were demanding of us if we wanted to have any hope of connecting with a big rainbow trout or grayling. I looked over and watched Bob's effortless casting—rod, line, boat, and Bob seemingly moving as one entity. Like all people with mastery of a skill, he makes it look easy and never seems to think about it. I, on the other hand, must remind myself to shoot a streamer and lob a bobber and nymph rig. I hope I can live long enough to begin casting like Bob does—you know . . . thoughtlessly.

According to Bob, the waters of the Agulowak were as high, fast, and cold as he had ever seen, and it was impacting the fish and fishing. This has been the pattern of my travels—every place is being impacted by human-induced climate change in one way or another. It may take the form of extreme heat, cold, flood, or drought, but the common theme is the word "extreme."

Over the next few hours, fish were caught. The day gave us a mix of Arctic char and grayling, but that big rainbow trout eluded us. It mattered not. I can't recall a more meaningful day of fishing or living in my relatively long lifetime. It was enough to be casting and occasionally catching beside my friend in the waters that he described as the home waters of his heart.

I came here seeking many things. I wanted to become engrossed in the adventure of flying over wilderness and landing on the water. I wanted to fish among the bears and moose and experience the eerie exhilaration of living beneath a midnight sun. I wanted to force myself once again out of my zones of comfort and familiarity by doing things I had never done before. I wanted to become captivated by these wild North American lands and waters that are the least impacted by humanity. But perhaps more than anything, I wanted to share these waters with my dear friend Bob and hopefully come to understand how and why they have captured the heart and imagination of a sixty-five-year-old boy from the American Midwest. And I did.

It was hard to believe that this was only my first day of living in and around Bristol Bay. It already felt like home. And I found myself feeling both elation and regret as our boat once again bounded across the waters of Lake Ageknagik toward the docks of Bristol Bay Lodge. I recalled my childlike joy at my brief inaugural flight the day prior, as I watched floatplanes approaching the lake, banking toward home, resting on the water, and taxiing toward the dock. I looked back at Ethan as he slowed the motor—allowing space for the planes to pass—wind in his young face and the expression of youth being bold in his choices and prudent in their choosing. I looked over at my friend Bob and noticed his expression of peace and place and belonging. And I looked toward the dock where Steve Laurant and his son Wyatt were standing at the ready to catch the ropes of an incoming floatplane, and toward the gazebo where my new friends Chelsey Faehrich, Stefanie Bollheimer, and Ellie Laurant stood smiling and waving as we stepped out of the boat and I gratefully accepted a glass of pinot noir from Chelsey. This is very much a family-and-friends operation, and I felt instantly at home among them. In fact, I miss them already.

My second day on the Wok found me paired up with my cabinmate and friend Steve Negaard, while our guide was a congenial and capable young man named Jack McGrain. Like me, Jack had Texan roots but had allowed the branches of his life's tree to grow toward the lights that called to him in Montana and Alaska. I remember those youthful days all too well, when I had few attachments and a plethora of desires. The first state of being is healthful, but the second can be problematic if not kept in perspective. In this late autumn of my life, "desire" has been replaced by curiosity—a much healthier approach since it can lead you to the same places but not leave you feeling as if you "missed out" as life happens to you while you plan. Some things I write in repetition, not because I have forgotten that I shared them before, but because I care about you enough to repeat them. In that spirit I will repeat: The root of all human suffering is desire and expectations left unfulfilled. It's best to just be curious and learn as you go—one cast at a time.

Steve and I were of like mind in that we hoped to focus on the big leopard rainbow trout of the Agulowak, with Arctic grayling and char being secondary and tertiary targets of our affection. Jack was all in and assured us that we'd catch some "big bows" before the day was out. We had both gotten the word that Jack was a bit of a specialist in catching the massive and stunningly beautiful rainbow trout of these waters, and because the sun was out and the rain was gone, we felt as bright and optimistic as any three "sports" (guided clients) crossing a wild Alaskan lake and ascending a free-flowing river might feel. Still, as we made our way upriver to the first fishing spot, I couldn't help but remind myself again that the root of all human suffering is desire and expectation. Some things run in a loop in my head—the product of decades of hard lessons learned.

As we bounced and wove around rocks and rapids, my eyes strained to absorb every moment and memory of this place and its place in my heart. I knew, even then, that I had fallen hopelessly in love with a wilderness that, left to its own and without the adaptations of cabin, lodge, and floatplanes, might kill me. A bald eagle flew overhead and landed on the top of a dagger-shaped spruce.

Jack swung the boat into position at the top of a long-curved pool that the local guides call "the aquarium." The reason for its name was obvious: It consisted of a relatively shallow bend in the river where the angler can see all the way to the bottom, which ranged from two to ten feet deep. The shoreline was dominated by alder trees with a phalanx of Sitka spruce behind them. There had been quite a few bear sightings along this shore, and I was eager to have one of my own. I was happy to learn that even though the guides saw brown bears almost every day, they never lost their sense of wonder and respect for these magnificent ancient beings. Neither have I.

Once again, due to the high, fast water we were forced to do the kind of fishing I enjoy least—nymphs under a bobber. In this case we were using a flesh fly with a nymph dropper. Steve and I began the repetitive dance of lobbing the rig out into the current, following the drift, watching the bobber, then repeating the sequence until the bobber bobbed. While we did this, Jack drifted us downcurrent until we came to the bottom of the pool, then fired up the motor and brought us back up to the top of the pool so we could do it all again. Methodical, repetitive, and somewhat meditative. Not exactly exciting, but it did allow for the low-energy thrill of anticipation.

One of the reasons why I'm not a big nymph rig fan is that it is methodical and repetitive, and to my mind takes less skill than other forms of fly angling. Often it seems that the most skillful part is being able to make a wide, loopy, lobbing cast so as not to tangle the leaders. The other reason I'm not a fan is that it can become so monotonous that you get lulled into a trance of complacency, and when the bobber finally bobs you might hesitate to make the strong downstream hookset that is required. And since you're mending the line upcurrent and setting the hook downcurrent, things can get messy.

Still, as a self-proclaimed "Imperfect Texan Buddha and Warrior-Poet," I should welcome the Zen-like qualities of repetition and the need to remain focused and in the moment. Like washing the dishes, I reminded myself that everything in life is an opportunity to practice being mindful and present. I cast again and watched the bobber drift. Perhaps I would get better at this if I wore crimson robes.

We made several passes through the aquarium before moving down to another bend in the river that up until this day had no given name that Jack was aware of. We were about to fix that situation . . . but I'm getting ahead of myself in telling this story, so let's start at the beginning. Steve and I both picked up a fish or two in the aquarium—a couple grayling and a single Arctic char for me—and experienced a few missed opportunities whenever we got complacent or just caught up in the scenery, which is easy to do. The fishing was wonderful, but the catching was slow. Jack was getting a bit frustrated because he really wanted to get us into some big rainbow trout, and we wanted that too. But the river decides what it will give you—water to drink or to drown in. So we kept casting and drifting until that moment when the river decided to give me a chance at a fish that led to this lovely little pool finally getting a name.

It happened on our second pass downcurrent, when I was casting into the water but mostly searching the coastline for bears. I really wanted to see one, and this was the area where a sow with cubs had been seen the day prior. So I'd cast and drift, looking down at my indicator and up for any indication of ursine activity, and that's when I noticed my bobber being pulled strongly downward, like the stock market and my retirement portfolio had recently done. I was so surprised that when I set the hook and realized that what I finally had on my line was a "big rainbow," I wasn't prepared for the trout to take all the slack line that draped below my line hand—all at once. The line ripped through the first crease in my middle finger, which was appropriate as I shouted an expletive, let go of the tension in the line, and lost the fish! Jack looked at me and said, "What happened?" I looked at him like a kid on Christmas Day and said, "line burn." We all broke out in laughter—Steve and Jack at the pathetic look on my battered face, and me in embarrassment. That's when my buddy Steve dubbed this previously unnamed section of water "the line burn pool." Of all the things to be known for, it had to be that.

Whenever I fall down, I get back up, and so we went back up to the top of the pool and began working it one last time. I felt determined to redeem myself for my sins. And that's about the time I looked up from my bobber and noticed something massive and primal standing in the water, staring at me. It was a huge brown bear with two fairly grown cubs.

I was thrilled. Not only did I finally see my first coastal brown bear in the wild, but I had good reason to suggest that we rename this alder-covered cove "the brown bear pool." Steve rejected that idea out of hand, laughed, and said, "It will always be the line burn pool . . . there's no escaping it." I may have jokingly used another expletive.

Around the time the sun appeared to be at its apex in the Alaskan sky, we pulled over in another cove and beached the boat next to two others where we met up with Ethan and Sam and several other anglers for a shore lunch of potatoes and fresh-caught sockeye salmon breaded in beer batter and fried over an open fire in a huge cast-iron skillet filled with oil. None of this was part of my heart-healthy diet, but I didn't have the heart to turn it down. It was perfect.

While the guides were cooking the shore lunch and the other anglers were sitting on logs telling stories, Steve and I kept fishing and I managed to catch another grayling. In the evening Jack worked hard to get us into rainbows and Steve finally caught a couple of nice-size fish, both grayling and rainbow trout, while I caught a beautiful little rainbow of about eight inches. I was grateful even if it was so small that it might have lived comfortably in a goldfish bowl. The big fish eluded me. Somehow, I managed to hook and lose another big 'bow. While I was catching fish as well as could be expected on a slow fishing day during the morning, in the afternoon, my mojo just wasn't working. It mattered not. We had a wonderful day together on the Wok. And since I had hopes but no expectations, I didn't suffer a bit—except for a little line burn. That's a small price to pay for moments and memories of laughter, joy, and gratitude among friends out in the Alaskan wilderness. Ain't life grand?

As I write this, I am sitting at my kitchen table in my humble house in my little Texas Hill Country town, watching birds at my feeder and being thoughtful that today is the seventh anniversary of my father's passing. I miss him every day. And that's the downside to loving anyone or anything or anyplace. Love always leads to loss and yet, like passing our DNA through time—love lives on. Love is timeless and immortal. I get the feeling that my father was with me on that first floatplane flight—smiling at the wonder and joy on my face. And I suspect he chuckled and recalled taking me fishing as a small boy as he watched me now, at almost sixty-three years of age, still having to untangle my line and try not to fall in the water. Old men are just young boys in well-worn bodies.

I remember the days when we all wrote letters to each other. We took the time and invested in our friendships—even if we were separated by miles and moments. And whenever we could, we got together and looked into each other's eyes and saw the soul behind those eyes—and valued that gift. We have been losing our way in these "modern times." We give up on relationships and purposely distance ourselves so that we can avoid the pain, when all the while we are avoiding the joy. It's not for me. I won't live a half-life. Instead, I will risk the hurt so that I don't risk living an illusion—a lie. And I am taking that risk with you now. I am taking you with me on this adventure in living—and dying. I want you to laugh beside me and perhaps even cry with me as we both love, embrace, and even mourn this beautiful living planet and the phenomenon we call "life." These pages that follow are my letters written to you, because no matter where, you are my friends—We Travel Together. We are all just *casting homeward.*

Chapter Three

Upnuk Lake by Floatplane, Alaska

Once you can accept the universe as matter expanding into nothing that is something, wearing stripes with plaid comes easy.

—Albert Einstein

All my life I've been told that I'm "too much." Even my own daughter has said, "Dad, sometimes you're just too much." Only during my time in the Marines did I seem to be just enough to be a part of a tribe that prides itself in being "more." But if I'm honest with myself (and I try to be), I was too much then as well. I'm not too sure that my fellow Marines pondered the deeper meaning of life as they cleaned their M16A1 rifles—as I did. Then again, perhaps they did. The specters of death and life create a wonderful gumbo of philosophical exploration and endless gratitude. And through the years I have concluded that the problem may not be that I am too much, but rather that humanity is often too little.

I suspect that my new friend T-Bird has been told more than a few times that he is "too much." And just thinking about this hurts my heart. After all, I know what it feels like to be on the outside of the universe looking in. I know what it feels like to want to fly tight loops around the universe while the choir sings, "Slow down. Steady as you go. Follow the leader." The thing is, no one is really leading; you must lead yourself. And just like T-Bird, I want to fly.

T-Bird, also known as "Free-Bird" or simply "Bird," has the little-known given name of Troy Adplamalp. It's a name that may take up space on his birth certificate, but it hardly describes his avian soul. To my mind, T-Bird is a perfect designation for this free-flying enigma who wears a de Havilland Beaver like a second skin. He began his love affair with flying as a chopper pilot in the U.S. Army and moved on to fix-winged aircraft shortly thereafter. And yes, to some people I guess that T-Bird might seem to be quite like me—"too much"—but I'll tell you this, my friends, I'd fly with him anywhere. He makes me smile.

What if we all had names that described us, or at least the person we wished to be? I wonder if we'd behave a little better if our actions decided our names. I think we'd all prefer to be known as "Wind in His Hair" or "Dances with Wolves" rather than "Lazy Lying Bastard" or "Mean-Spirited Narcissist." In a way, like *The Portrait of Dorian Gray*, our faces show who we've chosen to be in life—and it is a choice. Do the lines in our faces signify years of laughter or disdain? We are all just the product of our upbringing until the moment we choose to bring ourselves up to be more than a mere result or reflection. We can be the person of our choosing. I choose to be free.

T-Bird seemed to seek me out on that first day, and over an Alaskan amber brew he whispered in a conspiratorial tone, "Steve, you need to come out with me on a hop. I have lots of secret places to take off and land where the fishing is wonderful, and the scenery will blow your mind!" Needless to say, I was all in and so was my buddy and cabinmate, Steve Negaard. We had been assigned to the original cabin that started the Bristol Bay Lodge. It was called Nerka Cabin for the lake by that name and the fish it was named after—namely, sockeye salmon. We had our own separate rooms with en suite bathrooms, and shared a common living space that included a fridge full of brews, a fireplace full of wood, and a picture window that was filled with a view of Jackknife Mountain across the lake. If I could, I'd live there for the rest of my life—in the summer.

Steve Negaard and lodge owner Steve Laurent are best friends who have known each other since middle school. Now I counted them as my friends, too. I've never been anywhere where there was another "Steve," and now there were three of us! (We also had two guides named Jack and two named Ethan.) It didn't take long for my cabinmate and I to bond and find that we were perfectly matched as fishing buddies and unstoppable jokesters. In fact, laughter seemed to be the common denominator of our friendship, and at times we found ourselves laughing so hard that we had to stop fishing—and that's as good a reason to pause a cast as any I know.

The other way Steve and I matched up perfectly is our casual and relaxed way of dealing with the variables of bush plane flying—variables like staying airborne, not hitting a mountain, circling over bears, and "Can we actually get there from here?" Steve was kind enough to allow me to fly shotgun with T-Bird every time, which worked fine for him. While other travelers might lose their cookies during a series of tight loops and low landings, Steve was more likely to take a nap in the back of the plane while I sat up front completely enthralled with every magnificent maneuver and bit of aeronautic alchemy that T-Bird seemed to manage with ease. I love flying in small planes over wild landscapes. Also on this trip was a young man who T-Bird called "T-Rad." He was our guide for this adventure, and I found him to be professional, knowledgeable, confident, and quietly reserved. In short, I found myself in the best of company.

Almost as soon as the pontoons separated from the waters of Lake Ageknagik, I noticed that my buddy was off to slumberland in the back of the plane. That was about the time that T-Bird leaned over toward me and, as we both lifted our earmuffs, said, "Since it's just us, can I be myself?" I smiled and said, "I'd be disappointed if you weren't!" You see, with other guests Bird would have done everything in his power to ensure a smooth, straight, and to my mind, boring ride. But with me he was free to fly circles around wildlife and scenery so that I might take photographs or simply get a better look. We could be more carefree in the way we arrived at any destination. With me he knew it was absolutely fine to slide the plane down a side canyon or get a closer look at a bear standing in a river or just do a few loops around a particular mountain because it was beautiful and neither of us wanted to leave its snowy peaks quite yet. We were simpatico. We were truly alive. We were free.

Our flight path took us over the lakes of Aleknagik, Nerka, Nugakuk, Chikminuk, and, incongruently, a lake named "Beverley." I have no idea who she was, but naming a lake after her here felt like planting a rosebush in a wildflower meadow. It was just wrong. It felt unnatural and even intrusive. I much prefer the names given to these waters by the native Yupik people.

The Yupik are an indigenous people of southwestern Alaska and the far eastern Chukotsk Peninsula of Russia. They are believed to have crossed the Bering Land Bridge about ten thousand years ago, and their relatives still reside in eastern Siberia. They are related to the Inuit and speak a version of the Eskimo/Aleut family of languages. Culturally they are unique in that they name each newborn child after the last person who died in that community, so that both biological and cultural identity are passed on from one generation to the next.

It's sobering to think that when the Yupik first migrated into Alaska there were still woolly mammoths grazing on nearby St. Lawrence Island. When *Homo sapiens* arrived on the island, these ancient creatures were quickly hunted to extinction. All too often, the human act of discovery is an intrusive and destructive force.

As I looked down at the vast wilderness expanse of lakes, rivers, mountains, and bright green valleys of Wood-Tikchik State Park, I felt grateful for their protected status. I once wondered if I might be a "libertarian." But I discovered that I believe in the value of publicly protected lands, waters, and wildlife. I see the value of being a nation of public education where every child is exposed to the vastest array of ideas and perspectives. I see the urgency to come together and agree about placing boundaries and limitations on the impact we have on each other and the earth. I am a freethinker in a world that demands conformity. I have no politics—left or right. Instead, like the Yupik, I am a member of an undersized and underrepresented tribe. I am a naturalist and a warrior for peace.

We flew over snowcapped mountains for over an hour—T-Bird, T-Rad, me, and our sleepy friend in the back of the plane. All the while I could not take my eyes from the open window of the cockpit. Razor-sharp ridges of stony mountaintops rolled out one after another in a vision of light and dark green forest and tundra, blue-gray talus fields, and blindingly white snow. At one point I was transfixed by the sight of our shadow on the mountainsides. The floatplane drifted in reality and parody as a reflection on glacial blue-green waters and as a shadow on blue-gray stones. As we rounded a mountain and descended into a wet, boggy, deeply verdant valley, my eyes followed the river that formed it and ultimately spilled into our destination—Upnuk Lake.

T-Bird landed softly on the lake's surface and taxied over toward a point of willow-covered landscape that jutted out into the water like a natural dock. T-Rad had called out that we should "go to the point" but T-Bird replied, "I want to try this other point first." T-Bird was being T-Bird.

We set up a few fly rods with sinking lines and big Woolly Buggers, along with a few spinning rods with spoons that would sink quickly and contained a single barbless hook. Due to the depth of the water where the lake trout were holding and the distances we'd need to cast in order to be successful, I agreed to try the spinning outfit. To be clear, I would never use a rig with a treble hook, and we were all of like mind that any barbs must be smashed flat. I do not "rip lips" and would rather lose a fish than abuse a fish. All of us felt this way and for this, I was grateful.

Looking down at my wading-boot-clad feet I noticed that the stones that surrounded this lake were like none I've ever seen. Each one was a flattened, smooth, bluish oval that created a pattern similar to fish scales up to and into the clear water. I picked up one of these flat, smooth, oval stones and placed it in the pocket of my waders. I knew then that I would be leaving part of myself in this place, and I wanted to take part of this place with me when I had to leave.

The water was so clear that I had noticed, as we taxied toward shore, that I could see all the way to the bottom—some forty to sixty feet down. It was like a dream I once had where my Marine brother Dave visited me, several years after his death. (He does that from time to time.) We were in a boat on Lake Medina, which is really a reservoir in Texas. I was looking over the gunnel of the boat into the water, and it was so clear I could see the bottom of the same lake we once swam and fished in together—back when it was healthy and alive. In the dream, it was devoid of life. I asked Dave what it meant. He said, "Enjoy what you have now because before too long, this place as you remember it will be gone." I thought about that as I made my first cast. What if this lake were devoid of life? Would I still come here? I knew the answer. It's the same reason I no longer attend funerals.

It felt good to cast the big spinning rig after an adult lifetime of being a fly angler. It reminded me of where I started with my dad, not knowing any better back then and casting treble-hooked spinners—barbs and all. Making mistakes is not sinning as long as we learn from them and adapt to that learning. I would not trade my time with my dad casting to bass in our warm southern waters. Those moments were my gateway to my love of Nature, and in time I taught my dad to be more careful with the fish and to not kill snakes just because they were snakes and to help box turtles get safely across the road. Forgiveness is freedom. Freedom comes with responsibility.

T-Rad waded up to me and we stood together in peaceful silence as I cast and retrieved the silver-blue spoon and absorbed every ounce of the power and beauty of this place. Nothing needed to be said. The moment was perfect, just as it was. And that was when I noticed a stocky lake trout following my spinning spoon. Just as she came almost up to my submerged legs she struck, and I struck back, and then she went on a run out to deeper water while I kept the pressure on and hoped I was doing just enough to bring her in and not enough to break her off. I knew as she swam forcefully out toward deeper water that she was trying to get home. We had that in common.

T-Rad was at the ready with the net when I finally brought the laker in, and I was overjoyed as I briefly held her in my hands. As always, as I gently lowered this muscled, multicolored char back into the bluest of water, I whispered the heartfelt words, "thank you," and that's when she splashed and soaked my face and shoulders with the cold glacial waters that now dripped from my graying beard and off the brim of my hat. I laughed and looked at T-Rad and said, "She gave me exactly what I deserved." I was supremely happy.

Steve had been casting from a point a bit closer to the plane but had not yet caught a fish. I was hoping for him to catch one as I hoped for myself to catch another. But T-Bird suddenly announced that he thought we should move over to the other point, so we loaded into the plane, he started up the engine, and we taxied across the water to the place that T-Rad had indicated originally. It was a short trip.

Even though both of my new friends were technically on the job, working for Bristol Bay Lodge, I was a guest and not a client. You see, I have a rule that I try to adhere to: No one works for me. We are all just friends sharing a day together in a beautiful place as part of an amazing life. As is my way, I asked my new friends to fish with us, and they did. Before long we were four abreast, side by side, casting spoons and streamers into the sapphire-colored waters of Upnuk Lake. I will never forget those moments, which felt like a silent musical where sheer joy fills the air—infectiously.

What if the world were more like a musical? Could politicians and preachers induce us to hate and kill each other if they had to do so with song? Could we even manage to be angry at each other if we had to sing out our frustrations, or would it cause us to feel empathy for the singer and create a musical reply of regret and apology? I wonder if any political party or man-made religion could inflame fears and weaknesses if they had to communicate in poetry and song. And how much happier and more honest might this world be if we were compelled to sing out our true feelings rather than simply play the role we think we are expected to play? Sometimes I wish the world were more like a musical. Soldiers singing across a battlefield, trapped in their foxholes until they reach harmony. In a way, every time I cast my line, I am singing. Every time I release a fish and watch him swim home, I hear Pythagoras's music of the spheres.

I felt grateful to have caught and released my first lake trout, and I wanted to catch more. But I also wanted my buddy Steve to connect with one—and he did. In fact, he somehow found a sweet spot and began catching fish after fish, seemingly on every other cast. I was standing just fifty feet away and yet cast after cast I came up fishless. I was happy for my friend that he had caught a fish and even happier that he caught ten fish, but I must admit that as he grew closer to the twenty-fish mark I was beginning to wonder what I was doing wrong. That was about the time he released another fish, looked and me, licked his pointer finger, and made a checkmark in the air. We both laughed as I mimicked his antics in return—but I chose a different finger. If we can't make this a musical, it needs to be a comedy.

By this time T-Bird had stopped fishing and walked back to the floatplane. Steve and I had just remarked to each other how much we loved this place and never wanted to leave. We spoke in hushed tones about the absolute serenity and silence of these wilderness waters. And that's when we heard the voice of George Jones singing over the speaker that Bird had attached to his cell phone, and we realized that suddenly it *had* become a musical. You see, our pilot was singing along at the top of his voice. It was wonderfully surreal. I don't think my ears actually bled.

In between my buddy catching bunches of fish and me making endless casts and retrieves, I had three nice lake trout follow the spoon right up to my rod tip; one of them actually grabbed the lure at the last moment, but he was so close that I couldn't manage to set the hook before he let go. I was damn tempted to cast over to the honey hole that Steve had found, but friendship is more important than fishing, so this time instead of licking my finger I licked my wounds and walked back to the plane. It was then that I took the time to really look at my surroundings. It was peaceful again because Bird had turned off the music and was tinkering with fishing gear near the plane.

Just behind the floatplane's pontoons the soil was misshapen by the tracks of a bear and a single seemingly large moose. Next to the bear tracks I saw a twisted and sun-bleached length of driftwood, and out of a crack in the wood a bouquet of tiny white wildflowers was growing up toward the life-giving sunlight. Life really does find a way. Nature teaches the power of reliance and adaptation over rigidity. Fixed ideas and ideologies lead to extinction. Learning and adaptation lead to healthy growth.

A loon was calling from across the lake, and the soft sounds of wind and water filled my ears. It didn't sound lonely; it sounded at peace and at home. Looking into the deep green forest I wondered what the bear was doing now. Looking up into the surrounding mountains I wondered if I would ever see them again—except in memory. Life was beautiful, fish or no fish.

We were all having fun and I was quite pleased to catch my single fish, even if I was hopeful of connecting to more. Steve had caught quite a few, but he was openly sharing that his dream was to catch a really big lake trout, and he believed with all his being that the trout of his dreams was somewhere in the deepest part of this lake. He said, "It's too bad we don't have a boat out here so we could sit out in the middle and jig a lure over the bottom." I really wanted his dream to come true and so did Bird. And that's when Bird told us to get back in the plane and buckle up. We did, and he started the engine and taxied out to the middle of the lake, cut the engine, and said, "Be careful getting out on the pontoons. You can each take one pontoon and drop your spoon to the bottom and jig it as the slight current and the wind moves us across the lake." We did. This was so cool. We had a plane that was a boat. Adaptation is everything.

There was absolutely no risk here. T-Bird made sure that the engine was off, the plane was stable, and we were all wearing our water-activated personal flotation devices—but it felt adventurous nonetheless. And to be perfectly honest, unless we're talking about an Irish dance, jigging isn't my cup of tea, but Steve jigged gleefully with visions of a thirty-pound lake trout being released from the pontoon of a de Havilland Beaver dancing and singing in his head. I decided to lean against the open door of the plane for stability and then cast out as far as I could and retrieve my new silver and green spoon over the depths. It didn't take long before I had a strike and a hookset, and the fight was on!

I began applying side pressure on the fish in part because it was the right thing to do and also because the wing of the plane overhead precluded anything else. My rod was bent hard to the right and downward, and I could feel the throbbing weight of the laker as it bore down toward what it no doubt felt was a safer place in the lake and the farthest it could get away from me. T-Rad had walked the cable from one pontoon to the other (an impressive act in itself) while carrying a big net in one hand and leaning on the hull of the floatplane with the other. I began to gain some line and started to bring the big lake trout in closer to the rear of the plane. Just as I thought I might get my first look at him, the rod snapped back and the line went slack. I had put too much pressure on him and then somehow he had thrown the hook. I was briefly dejected by my mistake and the loss of the only fish I was ever likely to catch from a plane. But I remember my lessons, and while conjuring my best attempt at resilience and gratitude, I simply cast again while my friend jigged hopefully from the other side of the plane. We were having fun.

The problem with dreams is that we eventually wake up to this illusion we call "real life." And the problem with real life is that no matter how far we travel or how long the sun stays shining in the Alaskan summer sky, you eventually have to move on. I have to admit, I did not want to leave Upnuk Lake. While I was casting into its crystalline waters and marveling at the blue-green mountains that encapsulated it, for a moment I pictured a dreamlike cabin on its shores where I could live out the rest of my days. But then I woke from that dream and realized that by building that home and living here, I would change this place forever and not for the best. This is not my home. It is home to the lake trout and the bears and the willow trees and the ptarmigan. It does not belong to anyone, but part of me might belong to it.

We need wilderness. We need protected public lands not simply so we can use them for the extraction of timber, minerals, and fossil fuels, but simply because they have a right to exist and are the lifeblood of this magnificent planet. We have been given a plentiful paradise and we are turning it into a lifeless rock. I wonder, were humans once Martians? Where else might we have once lived and lost and left from?

Once again I found myself sitting in the cockpit of this plane that was as much a usable relic as I was. It was born just a few years before me. In a way, we were brothers, this plane and me. Like me, it had seen so many wondrous places and gone to the edges of the earth. Like me, it was struggling to remain intact as long as possible—to stave off the vagaries of entropy and remain airborne a little longer. All at once I was lifted from my trance, my eyes and my heart transfixed on this lake and its community of wildlands and wildlife. I heard T-Bird call out, "Clear!" and then "Contact!" as the engine rumbled back into service. I gathered every glorious sensation of the plane's movement across the water, the spray rushing past the edges of the pontoons, then the gentle, almost imperceptible feeling of separation from the water and connection to the sky.

I don't have a religion, but if I did, I guess I would be a Buddhist animist. Buddhist because I sense the value of letting go of attachments so that I can accept the vision of this magical place receding behind me. I see the value in understanding the impermanence of Nature and life. I use this understanding to help me choose wisely each day of my life. I choose love, kindness, courage, joy, forgiveness, and acceptance. I choose to be open-minded and keep learning and changing and growing—because when there is no more growth, there is no more life. And I choose to count my blessings and discount my burdens. Life is a series of choices. We can do nothing about what happens to us, but we can choose everything about how we will respond, not react.

I am an animist because I see and sense spiritual life in everything. I see a fish, deer, or tree as a living member of my community, not as a resource to be exploited. I see the earth as a living being complete with a lifeblood of water, air, and soil working in perfect unison, if we allow it. And I even see something soulful in a rock wall in the countryside, a bamboo fly rod in my hands, or a 1957 de Havilland Beaver that will never carry a more grateful angler than me. I think T-Bird feels the life in this old plane too. I'm not sure of his animist tendencies, but I am sure of one thing: If I were to give him a new name that reflects the offbeat, free-spirited, good-hearted misfit I enjoyed being with so much, I'd call him "Dances with Clouds." How cool is that?

Chapter Four

Floating on Alaska's Ungalithaluk River

The danger of civilization, of course, is that you will piss away your life on nonsense.

—Jim Harrison, *The Beast God Forgot to Invent*

I was flying shotgun next to T-Bird with my buddies Steve Negaard and Pue Nguyen in the back. I think Steve was napping, and I later discovered that Pue didn't enjoy Bird's aeronautic skills as much as I did. You see, on the way to Rainbo Camp my winged friend obliged my love of flying over wilderness by circling over mountains and around a huge bull moose and finally around the riverside camp, where we saw a big, beautiful blond bear running across a meadow while looking up at us—his unwitting tormentors. We weren't following him; he just happened to be in the meadow that was near the small lake on which we were about to land. But we did get so close to him that when he stopped and looked up, I could see the wind fluttering through the soft fur around his wide shoulders. And when we made eye contact there was recognition both of each other's aliveness and of the distance between our worlds. That distance was much more than the few meters from our wingtip to his uplifted ursine face. It was the distance between a creature who was completely connected to its environment and another who was desperately trying to find an impossible balance between wild connection and willful comfort.

There was a moment when I felt as if I had changed places with the bear, and suddenly I was looking up at the passing plane through his ancient eyes. I could almost feel the lingering questions: "What is this winged creature above me?" and "What does it mean to my future and my home?" With all that separates this wild apex predator and this half-feral airborne human, I felt that we shared some common history. It's as if there are memories of our ancestors hidden inside our genetic codes. Memories of past encounters before technology changed the most likely outcome. I can't explain it. It just feels more like history than fantasy. It feels as if we've met before.

T-Bird landed on the tiny lake and slid the plane around and up to the dock where we were greeted by our guides and new friends Ethan Stroebel and Cole Baldwin. We grabbed our dry bags and gear and began the long walk across the marsh along a wooden boardwalk that acted to ease the journey and protect the marshland in one altruistic action. Of course, the best way we can protect a place is to leave it wild and untouched by humanity, but since human activity has now impacted the entire planet from the bottom of the sea to the top of the stratosphere and from the Arctic to the Antarctic, sometimes the best we can do is to responsibly mitigate the impact of our existence. That is exactly what this walkway was intended to do. One of the many things that impressed me about Bristol Bay Lodge was that the owners and staff all seem to take great care to minimize their impact on the ecosystem and treat this landscape with all the love and respect it deserves. Even though they are only here during the summer season, for each of them, this place is home. With every passing day, it became my home too.

Rainbo Camp consists of a series of three Quonset-style tents for guests, two more for the guides and the cook, and another for the kitchen and dining area. A wooden deck with chairs and a firepit overlooks the Ungalithaluk River, which enters Bristol Bay just a few yards downriver from the camp. From the camp's shoreline I could see the Walrus Islands separating Togiak Bay and Bristol Bay, which in turn meanders into the Bering Sea. Just upriver is the confluence of the two branches of the Ungalithaluk, which have been given the Euro-American names of Rainbo and Wrong Turn.

A few piscatorial objects of our affection inhabit these tidal and fresh waters, including a smattering of king salmon at the end of their annual spawning run as well as incoming chum salmon that are actively spawning, resident Arctic grayling and rainbow trout, and anadromous Dolly Varden and Arctic char that are seeking to eat any salmon eggs they can find just downcurrent of the redds. In general the big king salmon were most likely to be found at the mouth of the river or up the main branch of the Rainbo. The big rainbows, grayling, and chum might be found on either branch, but the least explored and fished area was far up in the backwaters of the Wrong Turn River.

That night over dinner my new friend and guide Ethan Stroebel and I chatted about our options. In the morning, I could either do what my buddies Steve and Pue had chosen—namely, pursue big king salmon on the Rainbo River—or Ethan and I could take the boat up the Wrong Turn River, a journey he described as "sketchy." I liked the sound of doing something "sketchy" on this adventure, so I inquired about the true meaning of the word as he interpreted it.

"Well," he said, "I wouldn't normally offer this to a client, but you're our guest and not our client. And I don't often offer this because it means negotiating ever-changing narrow waters that will have a mix of rock-strewn rapids, fallen trees, overhanging brush, and shallows where we might need to get out and walk the boat over the gravel bottom. And then there are the bears, of which there are many. But if we can get way back in there where almost nobody ever goes, I think we might have a good chance of finding a really big rainbow. It's a low-probability but potentially high-reward trip. Considering what I've learned about your sense of adventure, I thought I should mention it as a possibility."

If he thought he had to sell me on this, he was quite mistaken—I was all in from the moment he described the wildness of the river and the potential adventure. It wasn't big fish that called to me, it was big adventure! And there was something special about exploring the "Wrong Turn" that appealed to my deeply adventurous and poetic soul. Ethan smiled as soon as I said, "There's no doubt in my mind. . . . I want us to explore the Wrong Turn!"

That night after dinner, I was the only one still awake, standing on the boardwalk and watching the sunset over the marsh, shortly after midnight. I just didn't feel like hitting the rack in broad daylight, and also, I really wanted to see *this* setting sun. I knew this was my one and only chance to see the ending of this day in my finite life, in this timeless place. And it was worth it. I stood there alone as the sky bled from bluish-purple into salmon-colored layers that melted into a sliver of bright yellow-white light just above the distant horizon. As the light grew ever more dim, I heard the calls of dowitchers in the marsh. They sounded lonely.

Gradually, the daylight faded and gave way to a raven-colored blanket that became studded, one by one, in long-ago memories of starlight. I wondered if the stars that once sent those luminescent photons toward my aging eyes still existed. I thought of how my words written within the pages of books were my best imitation of starlight. Someday somewhere, I hoped, someone might read my words and wonder if their creator still "lived"—perhaps in some other dimension. I wonder about that too.

It felt meaningful—standing there alone, in the dark silence. And it was about then as I walked back to my tent that I began to wonder what the bears were doing out there in the dark, and where exactly they were doing it. It's silly to feel safer when you zip up your tent and get under the mosquito net draped humorously over your cot. Still, whistling somehow gives comfort in the graveyard—as if whistling chases away zombies and tent zippers keep the bears away—and I will continue to whistle as I pull up that tent zipper.

The next morning, I was the only one awake and standing on the boardwalk to watch the sun rise over the distant hills. I was covered from head to toe in clothing including a mosquito net over my head and neoprene gloves on my hands. Even with all this protection the incessant drive that these little insects had to use the protein in my bloodstream to propel their sole effort to propagate their species seemed somewhat admirable. More than one died in the process. Natural selection was on full display, and I considered myself an agent of its mandates. It was a bloodbath battle—"Winner Takes Nothing."

After a quick breakfast of egg sandwiches and coffee we loaded up the boats, with Steve and Phu planning to go up the Rainbo branch with their guide Cole Baldwin in search of big king salmon while Ethan and I loaded into the other boat with visions of exploring a winding, narrow, and sparsely traveled stretch of wilderness. I felt all the excitement of a true adventure as Ethan strapped the Springfield 10mm pistol to his chest and said, "We're not taking any chances with the bears up there." I wholeheartedly approved of his prudent approach to fishing in brown bear country, but said, "If you get the firearm, I at least want the bear spray." He handed it to me, and I strapped the canister onto my wading belt. I've carried it before in Montana and southeastern Alaska, and even though to my mind it's the equivalent of a tent zipper, it too gives me a bit of false comfort, and the feeling that there is something I can do. Of course, the best thing we could do is use good sense by being vigilant and avoiding conflict with these incredible native creatures. I would not want to see an American wilderness without its apex predators, and places where we have removed them, for me, are no longer truly wild. They belong here, and we are the interlopers.

I wish I could honestly say that I'm a "wild man," but at best I'm a semi-feral traveler. And I sense there is a difference between a traveler and a tourist, but it might be only slight, and I might be deluding myself. The difference as I understand it is that a tourist goes somewhere and watches the goings-on as if looking in from the outside, while the traveler does their best to become part of the community they are visiting instead of apart from it. And honestly, while I might be a traveler in that I am seeking to immerse myself in the experience of Bristol Bay Lodge and Alaska, I can't say that I'm more than a tourist in the wilderness. After all, I am not inclined to remain outside with the bears, moose, wolves, and mosquitoes as my first choice. At the end of the day I'd rather have my friend Chelsey hand me a glass of wine as Stefanie informs me that today's salad comes with a pomegranate dressing over fresh greens with strawberries. So I guess in that way, I must sadly admit to myself and to you, I'm just a tourist in these most wild of landscapes. I may not be ready to run with the wolves, but I'd love to sit on the porch under a starlit sky and listen to their songs of unity and freedom. Perhaps I'm more feral than wild, but deep down inside I harbor a primal yearning to howl at the moon.

Ethan started the motor on the boat that would take us up the winding and remote Wrong Turn River—and hopefully back. Cole motored away in an identical johnboat except that his contained my friends Steve and Phu. We followed them upriver until the confluence where the Rainbo and the Wrong Turn became the main branch of the Ungalithaluk; at that point they turned left with the rising sun just over their right shoulders and we turned right with it warming the right side of my face. Ethan deftly steered the boat around the many turns, twists, and occasional oxbow loops of this mysterious water passage. We skimmed over shallows and skirted around small, rocky rapids and under overhanging willows, and no doubt passed more than a few unseen brown bears—just beyond the willows. And we kept winding back and back and farther back into increasingly wild country as the river grew narrower and the salmon rolled more frequently, as if they were joyful about coming to their ultimate end.

Could there be wizard-like wisdom in a coho? Do king salmon and sockeyes contemplate the circle of life? I doubt it. More likely their wisdom comes from being authentically wild and native. There is no false pretense in them, and nothing is held back. They're all in from the moment they breathe in their first gill-full of water. They are born to seek to survive long enough to return to the place where their parents gave their last measure. There, like so many generations before them, they pass on the messages of their ancestors and, in doing so, let go of their mere mortal existence. Take away the trappings of human fiction and there's little difference between us. We tell the same story whether in DNA or pixels.

Once we had traveled about as far back as we could, Ethan spun us around and cut the engine, dropped anchor, and said, "This looks like a good place for mousing." I was thrilled. The mousing process is simple: casting across and slightly upriver and allowing the mouse imitation to drift while raising the rod tip and ultimately stripping in the line in a slow, steady manner so that the "mouse" appears to be swimming and not jerking across the water. Ethan watched my first few attempts and said, "You want the nice V-shaped wake coming from the mouse's nose." It took me a few more practice casts to get it down, but soon it became second nature and I marveled at how much the bit of fur and fluff looked like a living, swimming mouse. If I was a monster rainbow trout, I'd eat it.

I was so impressed with the look of the water and generally pleased with the way the little imitation mouse was swimming that I was surprised that the smashing strike I anticipated never came. So we moved downstream to more likely water, and I kept casting my little rodent hopefully—to no avail. It was time to switch to a streamer.

There were allegedly "big fish" in this little backwater, so I was using an 8-weight rod and a big and somewhat clunky streamer. I had been working a bit of fast water to the port side of the boat when Ethan said, "Give the slack water under the willows a try." I did, and on the third cast and retrieve I saw a large, long shadow of a fish following the streamer almost to the boat, and then turning away and heading back into the depth of the slack water. Ethan saw the follow as well, and we were both excited as I cast under the branches once more. Neither of us could tell what kind of fish it was. I'd have to catch it to find out.

I cast again, but this time there was no follow on the retrieve. I tried once more, and with this retrieve I saw him coming. It all began like a reflection of the first time I saw him, a looming shadowy fish slowly following my dancing streamer until all at once the decision was made in his salmonid mind and he rose up and struck the fly savagely. Upon contact he turned like a Minnesota musky as if to carry the fly back to his submerged lair. I struck back with just as much vigor and the fight was on! When he turned I saw the flash of his dorsal fin, and it was then that I knew I had hooked into a large Arctic grayling. He led me out into the deeper water, and I fought to keep him out of the quicker currents where he might use them to his advantage. There was a nice splashing leap and a couple of rolling dives into the mid-river depths before I was able to work him in and raise his head so Ethan could scoop him into the big net we carried just for such occasions.

While I briefly held him in my hands before sliding him gracefully back home, I couldn't help but be taken by his wild beauty. Ethan seemed excited for me and said, "That's a grayling of a lifetime in these waters!" It was a nice big grayling, but I wasn't sure if it was a "grayling of a lifetime" or if that was one of those things that guides intuitively learn to say along with "good cast" and "don't worry about it . . . I love untangling lines." I really like Ethan and have faith in his honesty, but the little voice in the back of my head always puts "happy guide talk" in the same category as if some old girlfriend said you were the greatest lover she'd ever known, and that you had ruined her for all other men. It's not that you don't want it to be true about the fish or your masculine prowess—it's just that it seems too good to be true. Then again, when I got back to the lodge a day later, I found out that Ethan had already communicated with the other guides about the fish and sent a photo or two, so maybe I did catch the grayling of a lifetime—whatever that means.

This small subarctic river drainage is a remnant of what North America once had in abundance; it's a riverine wilderness. There are no roads here. There are no trails. There are no artificial lights. There isn't even a landing strip, which is another reason I adore floatplanes—no footprint.

Sometimes I wish I had the capacity to live that way—leaving no trace. Other times I tell myself that if I'm honest, that isn't exactly true. Like my ancestors I love to paint on the cave walls. The only difference between my ancestors' narcissism and my modern attempts to communicate beyond the grave is that while they used pictures, I use words. We both seek to find a way to give meaning to our lives and pass on whatever we've learned along the way. When I write, I hope to give something of value to whomever may choose to travel with me. I hope to pass on some sort of wisdom that has been passed on to me—like a moral genetic code or a song that outlasts the singer.

We kept working promising stretches of water as we moved ever closer to the sea. And in this case promising water wasn't just pools, pockets, and runs but also those areas just downcurrent of the many active chum salmon redds on either side of the river. Angling causes us to learn the habits and habitats of our intended target fish, and we knew that big rainbow trout and char might be predating on the salmon eggs that tumbled unfortunately out of their intended gravel-strewn nests.

As Ethan played the role of human anchor by wading in the river and walking the boat downcurrent as I cast, we noticed a large rolling mass of chum salmon that consisted of several tiger-striped males fighting over several big females. I began casting toward the places where the rainbow trout and char might be hiding and feeding on eggs. I wasn't targeting the spawning salmon, but it took no time at all for one to come charging out of the redd while angrily slashing at my fly. I set the hook, and the battle was on—and a battle is exactly what it was.

It was a big, heavy, highly irritated female chum that immediately took me into the depths of mid-river. Once I got her on the reel she went on a downstream run that took me into my backing as she leapt and sounded and rolled farther and farther away from my bent rod tip and aching arms. Ethan tried to pursue her with the boat, but the currents and shallows were too tricky, so in short order I was over the gunnel and wading downstream and across toward the shore. When I stepped onto shore I noticed some big bear tracks in the mud, but I took a quick glance around and continued on—I wasn't planning on losing this fish.

Over time and with effort, I managed to regain all my backing and some of my line, and I kept doing my best to work her into the shallows where Ethan waited with the big net. We got close, but as soon he made his move she decided to teach us a lesson in determination and the will to survive. In what felt like just seconds, I was looking at my rapidly diminishing backing and wondering if I could gain control before it came to an end. I chased her in big, clumsy, splashing steps and then along the gravel bars without a hint of thought about bears. And in between me losing and gaining line on my reel and the time I was finally able to wear her down, she managed to dive and leap and run and thrash with so much vigor that in the end I was left nearly breathless. Ethan and I both smiled in a mixture of relief and rejoicing as we looked into the net, where she began to regain her strength.

We kept her in the water as we unhooked the fly from her lip and carefully walked her back upstream as she revived in the quick, cold, rushing, oxygen-filled current. After taking off my gloves and holding her half-submerged in my wet hands, I watched as she slid back to the same spot in the river where we first met. And that's how I see it. We met briefly, and she taught me a few things about the will to live—even when that life was only a heartbeat away from a forever silence.

This mighty fish had survived predators and harsh conditions from the moment it was encapsulated in a gelatinous egg to this moment when she fought a true warrior's fight, in a battle that was preordained to end in death. (Isn't this our battle, too?) As I write this, I know she has completed her journey and is now part of the nutrient-rich waters and soils of the same wild landscape that gave her life. I will never forget her, and she will remain alive in my memory until my time comes to drift downstream.

Ethan and I continued to follow the Wrong Turn River as it twisted its way toward the Bering Sea. We watched the chum salmon as they shivered over shallows and rocked methodically in place beneath the surface of the deeper pools. I kept casting, as always, and in time I hooked another chum, this one a beautiful hook-jawed male. He did not fight as hard as she had. He seemed weary as I let him go, back into the waters I had pulled him from. He was handsome in his death colors of olive green across his back and the telltale rust-colored "tiger stripes" displayed vertically along his maroon-red sides. It occurred to me that those bars seemed not unlike those that might restrain a prisoner. And perhaps this fish, like me, is a prisoner of his own mortality. But even with the tragic image of drifting salmonid corpses, there was nobility in the fact that he and his mate had served their purpose. They had lived a full and fruitful life, and to me, there is something noble about spawning salmon. I hope I can accomplish as much.

I knew when we took the Wrong Turn that it would be a "low-probability but potentially high-reward trip." That is what I signed up for, and it is exactly what I received: From morning light to the evening's first sight of our camp, I had a blast! Over a full day of exploring and casting and catching a few fish, I never caught that "big rainbow." Yet I stood in the paw prints of a massive brown bear as I battled a big salmon who taught me a lot about living and dying—just when I thought I had already learned that lesson well. And I caught a big, beautiful Arctic grayling and later a smaller but just as lovely fish of the same species, both of them symbols of hope. Knowing that Arctic grayling are long-lived and slow-growing fish, I suspect the big guy was about thirty years old and the younger one just old enough to vote—if fish had a vote. And I did hook into and lose a couple of chances at rainbow trout, but it mattered not. Like I said, I had a blast.

As a writer, I'm supposed to be able to find the words to convey the vastness of this landscape and the deeper meaning of the journey, and, at least to some extent, I hope I have. But there can be no substitute for the visceral art of submersion in another world—you have to do it yourself. You have to be the one looking down from the cockpit of a floatplane as it begins its approach to a wilderness lake you will call home for the night. And then you need to know how it feels when you're looking up as that same plane leaves you behind, and you look at the weather that is accumulating on the distant horizon and wonder if it will return any time soon. It needs to be you who feels the spray of water over the gunnel of the boat or the way the hard, muscled bodies of salmon feel, just before you return them home. But it's not enough to go to a place like Alaska and treat it like a trip to Vegas where you count your "wins" and "losses" by the imaginary expectations you've created or assumed. It's not a numbers game of fish caught or the size of the fish—it's so much more. I do not fish to catch fish; I fish to be caught up in the act of being alive.

So, I will take one last cast in my effort to convey the vast beauty and intrinsic value of this landscape and the deeper meaning of the journey. This wild, wide-open, ruggedly beautiful, and potentially lethal landscape is a gift that is wrapped up in folds of snow, ice, water, stone, gravel, soil, trees, grasses, wildflowers, wild fish, and other wildlife. It reverberates in the echoes of Inuit communities who told stories and taught lessons to their progeny during long, dark winter days and bright summer nights. And it's here, in still largely intact, hopefully protected public land that through its sheer grandeur acts to remind us that we are all just stardust and seawater—and there is no such thing as someday. And I offer one last lesson that wilderness teaches. Sometimes the direction that everyone tells you is the "wrong turn" is exactly the right turn for you. If it feels right, do it. You can always reel in and cast again—until you can't.

When we got back to the camp, I was happy to find that Cole, Steve, and Phu had a nice day on the Rainbo River. Phu caught a few nice fish and Steve got three big king salmon, thus fulfilling his dreams for that day. As for me, it was bittersweet as I climbed into the cockpit of that glorious old de Havilland Beaver and saw the smiling face of my friend T-Bird. We all admitted to each other that it was going to feel blissful to have a warm shower and eat an incredible meal back at the lodge. And I will admit that it felt like coming home to taxi up to the docks and have my friend Chelsey hand me that glass of wine and ask me excitedly and earnestly about my adventure. But something lingered as I sat by the window of our cabin with a warm fire burning brightly and Jackknife Mountain reminding me that I was home. I think it was exactly that feeling of home that was gnawing at me. Part of me felt at home in the warm embrace of this cabin. And at least some of me longed to once again be in the middle of the wilderness where brown bears and golden sunsets spoke to the most feral part of my immortal soul. I guess I'm not such a tourist, after all.

Chapter Five

Floating on Alaska's Good News River

Be brave and walk through the country of your own wild heart.

Be gentle and know you know nothing. Be still. Listen. Keep walking.
—Mirabai Starr

I once met a man called "Musky Jesus." That wasn't his real name of course, but it might be a more accurate description of this quiet young man than the name he was given—Gabe. He got that nickname from his fellow fishing guides in Wisconsin and was not shy about sharing his discomfort with this honor, and I understood this, because he came across as a deeply humble, decent, and ethical being. They gave him this pseudonym because he seems to have an almost spiritual connection with musky and a supernatural ability to find them. But I see the name as fitting because of the deeper conversations we had both on the water and off. They were conversations that told me we shared a common inner silence that yearns for a better human world but lives within the one we've collectively created. Musky Jesus was born and raised "up north."

When I arrived in Birch Camp on the Good News River, I met three young men who seemed to personify much of what gives me a sliver of hope for the future of humanity and this dying planet. When we were on the water together and when we were sitting across the table from each other breaking bread and sipping wine, there was no difference in our ages or backgrounds or futures. We were all simply timeless friends living in the moment. It was beautiful. All three of these young men were born and raised "down south."

Like Rich, Jack, and Brother, I grew up in the American South. The southland I knew wasn't the angry, bigoted, fearful, or self-righteous place that I keep seeing in the media today. The southland I knew was kind, respectful, connected to Nature and natural living, bold yet humble. I'm not sure what has changed the most, my native homeland or my understanding of it and its people of every race and origin. I suspect it's a little of both, but mostly I grew up and grew in my understanding of human nature's darkness and light.

I now know that those darker aspects of humanity existed there when I was a child and that, sadly, they exist and even flourish in every place I've visited. Sometimes we are brave and wise, and other times, not so much. But my travels around the country and the world have taught me that no place and no people have a monopoly on ignorance or wisdom. And I also know that those things are not a reflection of the entirety of what is—or what can be. They are simply a snapshot of the loudest, most frightened corners of American culture.

Like me, most of the guides I met in Alaska were "sons of the South." They came from Arkansas, North Carolina, Georgia, Florida, Texas, and Alabama, and every one of them I found to be kind, honest, accepting, nonjudgmental, decent human beings. The three young men who greeted me and my buddy Steve Negaard were no exception. They personified what I knew growing up and I felt that we became friends for life, in an instant. I am grateful for the times we shared, and now I want to share them with you.

Rich Yancey and Jack Lawson were our guides on the Good News River, and Brother Swagler was so much more than the camp cook or even camp manager. He was (and is) a poet of sorts, not by what he writes but by how he lives. For me, if Gabe is "Musky Jesus" then Brother is "Good News Moses." I've always known what is all too often forgotten: North, South, East, and West, every land contains its prophets of peace and poetry. I wish the media would focus on those quiet, caring, accepting souls, no matter where their headwaters might surface.

The evening we arrived at camp after an hour-long ride in T-Bird's plane, Brother prepared a dinner that was as good as any I've ever had—and I've eaten a lot of good food. On a camp stove he prepared lamb chops with a side of absolutely amazing Tuscan-style pasta, and I had seconds of both—even though I had chosen to stop eating meat some six months prior in my desire to mitigate and even reverse my heart disease. But for this adventure I had decided to allow myself to experience it as it came—food, drink, fishing, and all.

While enjoying dinner we shared chilled white wine and warm conversation about the river, the region, the fishing, and the meaning of life. Afterward, we sat outside the dining tent where Steve and I sipped a bit more wine than perhaps we should have, as we laughed often and spoke of things both deep and meaningful more than once. It was a pleasant evening, and we watched the Good News slide by and the daylight grow soft and supple, like a child's blanket. It was comforting.

At one point, Steve, who describes himself as "a cobbler," seemed to become a bit melancholy as we both reminisced about the unfolding of our lives, and how life seems to happen to us while we plan. It was important to me that I stay with my friend, to listen and learn, and to simply be present. There wasn't anything he shared that I haven't felt myself. Like me, Steve is self-deprecating about his fishing prowess, but in truth he gets it done and done right, on so many levels. What's more important to me is he knows that fishing is about a lot more than catching fish. I guess that's why it was so bittersweet for me to know that we'd be going in opposite directions soon. I knew I'd miss our laughter and shared joy at simply being alive. I wouldn't trade a moment. I never want to lose the memory of that evening. These are the meaningful moments that lives are made of.

The next morning it was decided that Steve and Jack would go downstream to seek out more king salmon, while I chose to go as far upstream as possible in search of the Dolly Varden that were following the chub salmon in hope of feeding on their eggs. As I've shared in the past, I don't fish with "guides," I fish with friends. It just so happens that some of my newest as well as oldest friends are also guides. And so when Rich told me the night before that Brother was itching to do a little fishing, I invited him to join us. I am so grateful that he agreed. Fishing and friendship are two of the best medicines I know.

It was cool that morning and the mosquitoes were not as active as most other times. I stood outside the dining tent with my buddy Steve as Brother made us some amazing French press coffee. We all enjoyed the coffee with some egg sandwiches that were so delicious that just writing about them now makes me hungry. Life was good, indeed.

After breakfast we all met up on the edge of the river and loaded our gear and ourselves into the boats. My buddy Steve is a tall guy, and I noticed as he and Jack motored downriver to chase king salmon that he made the johnboat look small. It wasn't; my friend is just really tall. Then again, at only five foot six everyone seems tall to me, and yet somehow I've always felt tall—on the inside.

Rich, Brother, and I began our journey upcurrent along the Good News. Wet weather had come in during the night with a low cloud cover that seemed a bit worrisome for our planned flight out of camp later that day. But that was for later. For now, we were a band of brothers braving the elements in the middle of a vast wilderness.

It was raining softly as Rich navigated us around rocks and riffles and over rapids. I always enjoy the feel of raindrops as they contact the brim of my hat and roll with the rules of gravity off its edges and back into the river—homeward. Someday, I will make that same journey, just as the salmon that were spawning on either side of us were about to do. Going home comes naturally no matter what our plans may be for future travel and discovery. When it's time, the universe opens the screen door and calls out, "Come home." We all answer that call—ready or not.

I kept my eyes glued to the surrounding hills and tundra, hoping to see a moose, bear, or rare wolf. About a week prior, Jack and Rich had recorded a short video of a lone wolf loping along the very same hillside I was watching. Later, they found the paw prints of a single large wolf along the "Trail of Tears" that connects the floatplane landing site with the camp on the other side of a fairly steep bit of tundra. It's called the Trail of Tears because everything brought in must be loaded on the backs of the guides and camp manager and carried up and over that hill. Everything. Even boat motors.

I never saw that wolf or any moose or bears along those hills, although I did see two big brown bears while looking out the floatplane's cockpit window the day prior. But on this wet, wholly overcast morning, what I did see were many scores of chum salmon writhing in their redds and fighting for the opportunity to pass on their genetic memories to another generation. Behind these salmon we hoped were the char and rainbow trout that we would target while trying to avoid disturbing the many amorous chum of these waters. They had endured enough to reach this moment, and we wanted to leave them to it, while we kept their tormentors busy.

After a long, wet ride upriver we came to a confluence of two branches, and there we made our first play for the fish of our dreams. Rich jumped out of the boat and began acting as a human anchor. He walked the boat downstream toward some active redds as I stripped out some line and made my first cast. Almost instantly my indicator went under, and I set the hook! It was a nice Dolly Varden that gave me a brief but noble fight and seemed none the worse for it as I returned him to the river and watched him swim homeward.

I cast again and after about three drifts hooked into another Dolly. This one fought a bit more stubbornly and took me out into the depths of the river as I tried my best to raise his head and keep him from the quicker currents. Several times I got him closer to the boat, and yet each time as Rich would try to scoop him into the net, he'd go on another powerful run. This was an unbelievably powerful fish, and I felt every throb of his muscled body pulling line off my reel and time off the clock. Sometimes I wonder if I will ever stop catching fish and just begin watching fish—like a bird hunter turned bird watcher. But I know that I would miss the lessons they teach me when I am connected to them by a thin line and our mutual desire to continue living as long and as well as possible.

I landed that fish with Rich's help at the net, and we released him as quickly as possible once he had fully revived itself and could safely be sent back to the same waters he had chosen for himself. Whenever possible I try to return the fish to wherever I found them, because that spot in the river is where they chose to be. I know that there was something about the conditions of that location that the fish found to be "ideal." Who am I to decide otherwise? Even the fish I choose to kill and eat as "shore lunch" is treated with respect. It's important to remember where our food, water, and oxygen come from, so that we don't someday find ourselves without any of these gifts. Nothing in Nature is a given.

We had a big net in the boat. But Rich had also brought a little net that he said he bought for a couple of bucks at a shop in Arkansas. He seemed to love that little cheap net, and I understood that affection. Back home in my closet, I have an old cotton fishing vest that I never wear but never gave away. I'm thinking of taking it with me on a trip to Vermont where I intend to go "retro" and fish only with bamboo. It's an inexpensive vest of the old 1960s style with a "Trout Unlimited Life Member" patch on one of the pockets. I have "better" gear these days, but this is the vest I started out with as a much younger man who followed his heart, bought some inexpensive entry-level fly-fishing gear, and taught himself how to cast—sort of. So it has sentimental value that far outshines the couple of bucks I paid for it. I understood Rich enjoying the use of his tiny hardware store fishing net, but as a few fish popped off my line before we managed to net them, I began to wish he'd grab the big net and be a little less laid-back on those artful scoops he was capable of making. Don't get me wrong, every fish I lost was my fault. I guess what I'm sharing is I landed some and lost a few, including a really big char that broke my heart. Love often ends in heartbreak.

After catching and releasing a pretty consistent run of Dolly Varden, I implored Brother to put down his camera and pick up his fly rod. After all, we were friends now. I was happy when he did, and in no time we doubled on Dolly Varden with Rich having to work twice as hard to net them both. He was using the big net now.

We released our fish and stripped out some line for our next cast. All of us were smiling from ear to ear. Rich was standing in the water holding on to the boat when he said, "Bet you can't do it again." Brother and I looked at each other and said in unison, "Watch us!"

We cast simultaneously, and in a few seconds both of our indicators submerged, and we set the hooks on two equally big char! We were all laughing and overjoyed as our fish jumped and sounded in unison and Rich went in for the double scoop. I was a bit worried about a double scoop, but we maneuvered our fish together as if it were a dance. Rich went all in with the big net, and he did a magnificent job of lining up both fish. I was half-relieved as Brother's big char entered the net, and I tried to apply a bit more pressure and lift my fish's head. It was then that Rich went for it, and I saw her body halfway in and then watched in horror as she thrashed and flipped out of the net, off the hook, and back into the river.

They say (whoever "they" are) that triumph and tragedy come in sets of three, and this was my second heartbreaking moment that also contained a much more vivid feeling of joy and gratitude. We laughed at the wild ruckus we'd just experienced, and I patted Brother on the back for a job well done and told Rich that it was a noble attempt at a double, no matter the outcome. Still, I must admit that as an Imperfect Texan Buddha I have a long way to go before I reach enlightenment. I have sinned and must confess: I really wanted to hold that fish.

We worked our way slowly downriver, and both Brother and I caught fish. In time we came to a confluence where the river had been split by an island of willows and then reconnected in a perfect melding of currents and pools. Rich set the anchor and we took our lunch break on the edge of a gravel bar covered in brightly colored fireweed and ominous-looking brown bear tracks. The bears knew what we suspected—this place was fishy!

As we sat there eating our lunch and drinking our libations, I couldn't take my eyes off that fishy-looking water, and I licked my lips like a wolf at a shepherd convention. I did take the time to pick my head up and take in the magnificent wild beauty of this place. Along our starboard bow was the line of tundra-covered hills where the wolf had been sighted recently and bears sighted many times before. Between us and the hills were thickets of willows, and to our port were more willows that no doubt contained more than a few big bruins. The importance of apex predators cannot be overstated. And as a once avid hunter and still rabid conservationist and naturalist, I know that human hunters cannot take the place of natural native predators. We can supplement their positive impact, but not supplant it. Wildlife management includes the predation of the weakest herbivores, not just those with the "biggest rack." Natural predation keeps wildlife mobile and helps protect the riparian habitat that the fish we love need in order to live.

While we were eating, we studied the river and discussed our plan of attack. I wanted all three of us to fish in this spot, so we devised a plan to do that. We'd take turns, and since I was the "guest" Rich insisted that I go first, and frankly, I did not put up much of a protest. After all, they lived here, and this was probably the one and only time I'd ever get to experience the Good News—but who knows? Life's full of surprises, and many of them are wonderful.

We had decided to cross the river and approach the pools from the gravel bar on the opposite side. Before doing so, we each took turns walking in the tracks of the bear up to the edge of the willows, marking our territory so that Mr. Bear would know that "Kilgore was here." After all, we were not acting like tourists. We were completely immersed in this wilderness, soaking in its rain and river, itching and slapping at its malicious mosquitoes, and breathing in the oxygen from its willows as we gave them carbon dioxide in return. We were wild men, and together we loved every moment of it. In fact, as much as I wanted to make the flight back to the lodge that evening so I could share my last day with my buddy Bob White, I also did not want this day to end. I had felt the call of the wild and it felt natural. The rain came down a little harder and I cared not a whit. Wild men of the world—unite!

Once across the river I began casting and drifting the egg fly through the main current, allowing it to drift for as long as I could. After a few drifts I saw the bobber bob and I set the hook on what turned out to be a bright and fresh-looking king salmon jack. My second drift produced another king salmon jack, a little bigger than the first. Both fish were in the twenty- to twenty-four-inch range, which seems about average for these impetuous teenagers that just can't wait to return to fresh water. That impatience will cost them a few years off their life span. Rich said that he was seeing more and more jack-size males and fewer full-size males returning to the Good News River each year. Neither of us knew what that might mean for the future of chinook salmon, but it doesn't sound like good news.

After I caught the two jack-size kings, both Brother and Rich took a turn, and both pulled a char out of the pool. Then it was my turn again, so I targeted the main current once more and let the fly drift in what Rich dubbed "an epic drift." And it must have been epic, because at the end of it the indicator sank abruptly, I set the hook soundly, and the biggest Dolly Varden I've ever seen leapt into the air and landed permanently in my fondest memories.

If the drift was "epic," then so was the battle. She pulled me downriver and alternated between bearing down toward the bottom and leaping up toward the gray metal skies. I did everything I could to hold her until I made the fatal mistake of allowing her to get under a half-submerged tree, and in that instant of eternal regret, she broke the tippet and was free. Merde!

We moaned and laughed in unison as I stood there with a slack line and a comical look on my battered old face. I always try to roll with the river and rhythms of life, but I must admit, I will never forget the moment when that impressive fish first leapt into the air—or the moment I made the mistake of allowing her too much freedom. The end result was the same broken tippet I was trying to avoid. When things come unraveled in fishing or in life, the only thing we can do is learn, adapt, and cast again. No matter what—we keep Casting Forward!

While I was fighting that fish, Rich bestowed upon it the name of "Big Jimbo." After I lost it I handed the rod to him and said, "Why don't you try to remedy my mistakes by landing Big Jimbo?" Both Rich and Brother gave it a go, but Big Jimbo was wise to us now. Clever fishy.

While we were making our way ever closer to camp, the weather was rolling in with more malice on its face. It felt like the earth was frowning. The sky became a low-flung blanket of clouds that continued to drop rain and drizzle on our heads at increasing intervals, and the visibility was getting quite iffy for a plausible extraction by my friend T-Bird. As I said before, I was loving every moment of my time on these waters and with my new friends—but I wanted to get back to the lodge that night so I could share my final day on the Agulowak with Bob. There was nothing we could do about the weather, so we continued to fish while Rich communicated intermittently with T-Bird via satellite phone.

I should mention that each Bristol Bay Lodge camp is equipped with first-aid gear, backup food and water rations, and satellite communications complete with radio-like phones that accompany us into the wilderness. Each phone has a messaging system and an emergency button that sends out a beacon to surrounding potential rescuers that gives your exact location and indicates that you require immediate assistance. And each floatplane is equipped with first-aid and survival gear. If you go out into the Alaskan wilderness, do so with an outfit like Bristol Bay Lodge, which takes prudent precautions and puts visitor safety at a premium. Enough said.

We had time. That was one thing the lowering clouds and damp weather gave us. And for the most part I relaxed with it, knowing that what will be will be, and besides, if anyone could come get us out of this soup it would be the "Bird Man." So Brother and I fished, and we both caught a few more char and I caught two nice Arctic grayling, one of substantial size and both incredibly beautiful. Still, that big Alaskan leopard rainbow continued to elude me. And that's when it happened. Brother's rod tip bent, he set the hook, and in short order my friend caught and landed an absolutely beautiful, big Alaskan leopard rainbow trout. I was so happy for him—truly. I barely even thought about how close our indicators were drifting in the same current the moment his bobber plunged beneath the surface and mine sat up top looking round and reticent and even slightly sarcastic. Well, at least I *saw* a big rainbow trout.

When we rounded the last bend in the river and saw the domed tents of Birch Camp, it felt like coming home but also like the start of a journey home. Life's funny like that. And I'm still trying to get my heart, soul, and mind around this idea of "Home." What is it? Where is it? How do we know when we are home? Is it somewhere we arrive at or something we carry with us, like a turtle's shell or favorite old fishing vest? I don't know, but I'm looking forward to finding out and taking you with me on this adventure in learning about and loving life.

⁓

We stood there together at the end of the Trail of Tears with all our gear as we listened for the hopeful drone of an aging yet trustworthy floatplane being piloted by the equally attributed and seemingly unflappable T-Bird. The skeeters had regrouped and were launching another assault on my head and neck about the time I heard the sweet sound of a bush plane engine and soon after watched it touch down on the lake and taxi up to us as if it were the easiest thing in the world to do. But let me tell you, the weather was still precarious, so I knew this wasn't easy.

I jumped into the copilot seat and Steve got in back, stretched out his long legs, and may have been drifting off to slumberland before we even got airborne. I'm not sure, but I am sure that it's nice to travel with a relaxed frequent flyer. After takeoff, Bird leaned over and said, "We're going to be flying sort of low and be creative as we work our way under this weather and over to the lodge." I smiled. Maybe it's ignorance or maybe it's my Zen-like connection with the universe, but I wasn't the least bit concerned and was actually enjoying every moment of it. It was probably ignorance; I'm still working on my Zen.

For me, it was a beautiful and memorable journey. I relive it again and again in my mind and wish I could *do* it again. Rain was smacking against the windshield as Bird flew us low and carefully along a river basin and between the mountains and extinct volcanoes. I had my eyes fixed on the river below, and at one point I looked down to make eye contact with the largest brown bear I'd ever seen, as he looked up at us—at me. Even from the window of the airplane I felt as if we were occupying the same space and time. He was standing very much alone in the middle of the river looking up at us as we passed over him. Soon we'd be gone. Out of sight and out of mind, I suppose. And I know it's absurd, but I must admit, I want him to remember me.

You see, I will never forget him. When we made eye contact, I felt his aloneness, a feeling I know all too well. Perhaps I simply imagined that he might feel as I would down there standing in the rain. I told you, it's absurd. Still, it's how I felt.

It took a little over an hour for us to break through the clouds and bank over Lake Ageknagik as Bird pumped the handle for the flaps and adjusted the trim, propeller speed, throttle, and mixture to "full rich" so that we landed ever so softly on the water. I watched as Bird worked the pedals to the pontoon rudders as we taxied toward the lodge's dock. I looked over and said, "Nice flight!" Bird looked back, smiled, and said, "Thank God." We laughed.

I could see Steve Laurant and his son Wyatt waiting to catch the lines and bring us safely against the dock's edge. We opened the rear door of the plane and Steve said, "Welcome back!" but in my mind I heard the words "welcome home." It's not the same thing.

Bob White, my soul brother, was also waiting on the dock. It was good to see him, and when I got out of the plane we smiled and I gave him a hug. He didn't seem to mind. Bob's cool like that. As I grabbed my dry bag from the cargo hold and walked through the drizzle to the gazebo at the end of the dock, it was so nice to see my now dear friend Chelsey Faehrich as she handed me a glass of pinot noir and gave me a hug and a smile. I hugged her back. She didn't seem to mind. Chelsey is cool like that, too.

When I returned to the cabin, I found a roaring fire ablaze in the fireplace. After a warm shower that felt like liquid heaven and a clean set of dry clothes that felt just as good, I walked up to the lodge from my cabin where I got to visit with Steve and Bob and my other new friend Stefanie Bollheimer. She's the head of house who manages everything from the way the meals unfold to the way your laundry is folded, and so much more. She also manages moments, and one of them included me. After the pouring of wine and just before dinner, she asked me to read an excerpt from one of my books, which I did, and everyone was polite and kind and no one nodded off—so it went well.

After my reading I sat down next to Bob, and we shared the stories of our day. He had been investing his time working on a new painting of the rainbow trout I had never caught. You already know what I did during my day of adventure fishing with Rich Yancey and Brother Swagler. But what I didn't know until that moment were two important realizations. First, that it was Bob who built that welcoming fire in my cabin's fireplace. He said, "Knew you'd had a cold, wet day out there and thought it might be nice to come home to a warm fire in the fireplace." And second, that he was right about more aspects of his words and actions than he knew. It was indeed nice to find that fire alight in our cabin. And I had, in fact, come "Home." How cool is that? Finding home, anywhere, is the best good news.

Chapter Six

Alaska Epilogue: Return to the Agulowak

Every day is a journey, and the journey itself is home.

—Bashô

The week passed by like an unfolding dream—the kind that you wake from in the predawn darkness and wonder if you were dreaming or if this was simply a memory of another life that was lived while sleeping. Who is to say that our reality is not a dream? Looking back into my half-awake memories, I recall catching and releasing many salmon, trout, char, and grayling. I remember the sights and sounds of bears in the forest and loons on the water. And now, as I realize that I am back home in Texas, I lie there alone but not lonely, recalling my last day of fishing in the Wood River drainage of Alaska's Bristol Bay. It is a bittersweet recollection, indeed.

Time (whatever time is) had rolled out in front of me like dice lying motionless on a table. Rain fell across the Agulowak in soft sheets of mixed drizzles and downpours. It was cold but not unbearably so. Bob and I loaded into the boat, and our guiding-friend Andrew Tartaglio began motoring us through the chop, around the bends, and past the bears to our final destination, Grayling Island. Almost magically, I had come full circle.

Before our rendezvous with that spot along the Agulowak, where the unusually high and fast water was allowing the fish of our dreams to swim among the branches of alders and willows, Andrew took us up to "the aquarium," where Jack McGrain had taken Steve Negaard and me in our search for a big Alaskan rainbow trout. I had fond memories of this pretty little drift where you had the impression that every fish in the river could be seen through the clear, shallow water. But that was only an impression, not reality. What could be plainly seen were the many wounded sockeye salmon, locally called "cheese heads" because of the horrible scarring and infection many of them had due to their previous narrow escape from one of the commercial fishing nets that ply the waters of Bristol Bay. Any one of these salmon could have ended up on my dinner plate back home, so I don't want to be a hypocrite by decrying the impacts of legal commercial fishing. We are predators. But it's still sad to see these noble creatures so battered and broken as they swim relentlessly toward their destiny. They have traveled thousands of miles to come home one last time. I guess I'm on the same journey, battle scars and all.

The river was running high and fast, so as much as I yearned to fish a dry fly for rainbows, we were relegated to watching a bobber that was suspended over a flesh fly with a nymph dropper. This would be my last chance to connect with that imaginary aged-warrior rainbow trout. So I played along and lobbed the whole complicated contraption, while mending and mending and mending the line, for the longest drift I could manage. And it felt like a win that I was able to fish the entire length of the pool without tangling the leader or hooking an anchor rope. It was about then that I looked over at Bob and watched him make cast after cast that seemed both effortless and the epitome of perfection. Mastery is a beautiful thing to witness and an elusive quality to embody. My friend is as much a master of the fly as he is with a palette full of paint, a brush, and a canvas. He is also one of the best men I've ever known. I'm so grateful to be Bob White's friend.

A few days prior there had been several sightings of a coastal brown bear with a couple of cubs along this shoreline, and even though I had seen quite a few bears on this trip, I was scanning the alders and spruce trees in between casting and mending and watching the indicator floating without disturbance or hesitation. When my bobber did finally bob, I missed the hookset because I was looking for bears instead of "hooking" for fish.

The first time I came to Alaska, it was in the Tongass National Forest in southeast Alaska along the Inner Passage. A large, healthy population of coastal brown bears live there, and I kept a wary eye out the entire time I walked, waded, clambered, climbed, and cast through those primeval temperate rain forests. During that time I saw a multitude of bear tracks and gutted salmon and came so close to one in heavy cover that I could smell him, but I never saw one. For some reason I can't explain, I've always had a complex relationship with bears—one of admiration and primal trepidation. This time, I arrived with the desire to face my trepidation and grow my admiration. I managed both. I saw many massive brown bears on this adventure. Most were seen from the cockpit of a floatplane, but a few were fishing the same river I was. I noticed that I was transformed by this experience. I was no longer afraid, but simply respectful and deeply awestruck by their presence. It felt good to turn this corner.

Even from the cockpit, whenever we made eye contact, I saw the spirits of these great creatures—truly wild, truly free, unlike me. I am deeply grateful that there are places where bears rule and humans are watchful. I wouldn't want to meet one of these majestic bruins while walking alone along a rain forest trail or tundra hillside, but that is fine because I won't be walking alone in the land of the brown bear. My life tends this way. I am a warrior who detests war. I am a fighter who avoids a fight. I am a dreamer who stands firmly on the realities of this precious planet.

After a few passes with neither of us catching a fish, we decided to move to the place I most wanted to go on my final day at Bristol Bay Lodge—Grayling Island. I wanted to complete the circle with my last cast landing where my first cast had gone. Of all the fish I had hoped to connect with in Alaska, the Arctic grayling was the one I most wanted to hold in my hand and slide back into his native waters. I guess I wanted to send him homeward in the hope that in doing so, I might realize what home is for me. Or perhaps by connecting with this sacred fish and returning him to these hallowed waters, I might leave a bit of my soul here, rising toward the sunlight beside him and swimming among the river of raindrops. Epiphanies reveal themselves to the quiet mind, and my mind grows most quiet on moving water and in the stillness of the stars. Understanding is a process of discovery and revelation. I'm still casting. I'm still looking up into the darkness, while focused on the light.

When we motored up to Grayling Island, Andrew secured the boat in the quick, cold, current—first with the inanimate hunk of steel we call an anchor but then with his own fully animated being. Andrew is a former Penn State football team offensive lineman, and as such he was able to stabilize the craft in waters that would have left me struggling or helpless. We could see the obvious places to cast and drift our offerings, and occasionally we saw the fish. And we cast to them again and again, trying to entice them to either take the dry or the nymph, but we weren't listening to what the rising fish were offering to teach us. Learning requires open eyes, ears, and minds. It was only a matter of good fortune that we discovered the key that opened their doors.

Andrew acted as the human anchor and began moving us slowly into a new position along the edge of a community of half-submerged willows. Bob and I began casting a "squirmy wormy" that was suspended about six feet below a brightly colored indicator with a small nymph dropper about ten inches below the wiggling top fly. We could see some nice-size grayling rising up and taking something just beneath the surface, but they ignored our offerings until Bob allowed the whole contraption to briefly drift behind the boat and a grayling grabbed the "emerging" nymph. It was a gorgeous fish. He caught another that way, and I did the same.

Even though Andrew was technically our guide, I assured him, "We're just friends going fishing, so please feel free to fish." He did, and in short order he too caught a grayling while Bob and I watched and cheered for him and for the fish. It's not a competition. It's all about connections between people, places, and other living beings.

When the fish or the river tugs, we tug back. When we pull on a fish or push on the river, they pull and push against us. When they do, it reminds us that they too are alive with an innate desire to live or flow as long as allowed. If all you do as an angler is count the number of fish caught or record the length and weight of the biggest fish, you are missing the point of the entire experience. You might as well be on one of those stupid television shows where people get thirty seconds to fill their shopping cart with as much loot as they can manage. Connecting to a fish or a river is about the connection and the release. It's about being alive while practicing an ancient art, however modified by technology. It's about understanding each other and, therefore, our own mortal lives. It's all just tapping on the prison wall. Can you hear it?

If there is one thing that I most want to share with you through my humble words, it is that we have the life of our choosing. We can be aware, present, grateful, hopeful, and in love with life and living, or sleepwalk through a pseudo-life where what is "important to you" and what you "believe" is determined by what you have passively accepted from parents, peers, preachers, teachers, politicians, social media, cable news, and this opaque amalgam of misguided misperceptions we call "society." One path leads us to a fully engaged and meaningful life, and the other fashions us as a reflection of past lives that were equally passive and meaningless.

The world is full of imitations, but authenticity is what really matters. Just ask any really big trout. Nine out of ten old and wise Arctic grayling agree. No imitation is as good as the real thing. (I think I saw that on the internet, so it's got to be true, right?)

I am forever a deeply grateful man. I am grateful for the many souls I've connected with in ways that matter. I'm grateful for the many joys and moments of laughter and love we've shared. And I'm grateful for my bittersweet memories of the songs we crafted together that still play inside me—long after the music ceased playing live and in person. I am grateful for sunshine and starlight and raindrops and wildflowers that stubbornly grow from the drought-stricken cracks of sunbaked soil. I'm grateful for every fish I've ever caught and returned—and every one that eluded me. They all held lessons that needed to be learned. Nature teaches me everything I need to know.

After each of us managed to connect with a few pretty Arctic grayling and enjoyed sharing the simple serenity of a soft Alaskan morning, Andrew asked what we wanted to target next. But Bob and I were targeting the same thing—enjoying the day together on the water and off, in conversation and in silence. We had already decided that our perfect day today was one where we let it flow easily and naturally. I wanted Bob to be with me as I made the circle back to Grayling Island where I cast my first line into these hallowed waters and where I connected with my first mystical multicolored fish. And we both wanted to share a part of the day exploring and expressing our personal art forms together back at the lodge, with Bob painting as I wrote. This is the true joy of authentic friendship. There are many ways to cast our lines together—in solitude.

It's the same when we fish together. We stand in the same river casting to the places that swirl and tumble before our open eyes, perhaps only speaking occasionally. Often, there is no need for words between true friends. The silences speak as well as those times when we fill each moment with the sound of our voices. It is in silence that we cast and connect and say farewell to both fish and friends. Life is often like an eternity loop. It unfolds and folds on itself. It renews like a soft Alaskan rain. It changes but remains the same at its essence.

Andrew seemed to understand and resonate with our laid-back approach to this fishing trip. Fishing isn't just about catching fish, just as God doesn't live in a steepled building. Fishing is about connection to other souls, which is the same as connecting to our own true self. I think he knew that Bob and I had caught as many fish as we might ever need to catch in a lifetime. We both understand that the value is in the entirety of the experience—from first cast to last. In many ways, fishing and life are the same.

For me, the Native American view is the one that resonates with my own intuitive understanding. Everything is "alive." It's just that some souls are living as humans, and others are bears, salmon, birds, or even rivers. We travel together, each interdependent, while wrapped in the illusion of separation. The riffles, runs, and pools contain the river. The river contains riffles, runs, and pools. They are one.

It amazes me how quickly things come to an end. This is true with fishing adventures and lifetimes. This river flowed just as freely long before I ever took my first breath, and it will continue to flow long after I take my last. The grayling and I share the same finite fate and infinite future. And in the end, there is no end. There is only the mortal cycle of life and death and the immortal circle of timeless souls swimming and soaring through the never-ending universe.

The next morning I sipped coffee and stared wistfully at Jackknife Mountain. I knew that I would miss this place and these people. I walked out on the dock and jumped up into the copilot seat of the Beaver, and soon heard Ron calling out, "clear . . . contact!" as the engine of the floatplane came alive, one final time.

Only Africa has touched my heart with as much depth as Alaska. I remember back then looking out the window of the small commuter airplane that carried me over the Kalahari Desert, and as it did, I tried to burn the images of vast savannas, mopane forests, wilderness, and wildlife so deeply into my memories that I might never lose them—even after death.

I wonder if love can outlive our lifetime. Will my soul recall those people and places I love in this life when I move on to the next? I sometimes feel this is true, like those times when you meet someone for the first time of this lifetime, and you immediately think, "I know you." Whenever this happens, there's that lingering feeling that we have some unseeable history with the person's soul, and that if you could only remember it, you might recall the biological avatars that you both once inhabited in another distant time and place.

As the mid-century floatplane rose up from the waters of Lake Ageknagik and rounded Jackknife Mountain only to alight again, ever so softly, on the lake—I once again embedded the sights, sounds, and sensations of this place and time into my memory. As the plane floated upward I looked downward to see my new friends, looking up and waving with smiles on their faces and happiness in their hearts. I wondered if we'd ever see each other again. And in that moment I thought of my dear friend Bob, his wife Lisa, and his entire beautiful family, and I wondered, "Why is it that the people I love and love to be with the most always seem to live so far away?" It felt a little sad. It felt wrong. It felt like something I needed to act on.

Living has little to do with biology and much to do with the ghost within this machine we call "us." So many people are the walking dead. But now as the sunrise comes to my beloved Texas Hill Country home, my mind lives in the here and now, and also in the then and there. Every now becomes a memory—in an instant. Life is beautiful and dreams do come true. We simply must hold in our hearts and minds the image of what matters most—and keep casting homeward. We must all choose to live many lives—now.

Part II

Montana's Paradise Valley and Beyond

montana - Bob White

Chapter Seven

Wading DePuy Spring Creek, Montana

Nature itself is the best physician.

—Hippocrates

The first time I ever laid eyes on the Big Sky of Montana was shortly after I was honorably discharged from the Marines in 1985. My new wife and I had decided that we wanted to live someplace with more nature than neighbors, someplace where I could exercise my western-leaning soul and we might find our "one true home." Together we imagined a cabin in the mountains where we'd wake to birdsong while watching as mornings were born and evenings melted into starry nights. And we did the kind of things a young couple might do to make these youthful dreams transform into our life's reality. I enrolled in the wildlife biology program at the University of Montana, and we moved—sight unseen—to Missoula, where we rented a cabin on an 850-acre ranch in Rattlesnake Canyon.

It all felt quite dreamlike at first, until the realities of living in a place where at that time jobs were hard to come by and making a living was almost impossible. I spent almost every dime I'd saved while living off my hazardous duty pay in the Marines. We ended up leaving Montana nearly broke and completely homeless. It was a tough lesson to learn and a hard fall to bounce back from, but we did. No matter how hard I tried, Montana didn't become our home—back then. Who knows what the future holds, or even if there is a future to be held?

I guess I've been haunted by that quest, and now, many decades later, the answer to the question "Where or how do I find home?" still lingers unresolved. There was a time when I felt at home in the slash pines and palmettos of "Old Florida." But people fleeing the world they'd created somewhere else brought that world with them, and in short order all that I'd ever loved about my original homeland had vanished with the pines, palmettos, and panthers. And now my beloved Texas Hill Country has been "discovered" and is in danger of being "loved" to death in quite the same manner. Developers are destroyers, manufacturing soulless containers instead of vibrant soulful places for people and other living beings. Why? We can do so much better.

What to do? Do we stand and fight for the forests, hills, canyons, deserts, wetlands, grasslands, waters, wildlife, and climate of the places we call home? How do we win these battles when our adversaries have the money and therefore the power to pursue profit over prudence in this almost religiously capitalist world? And what about the impact of human-caused climate change with its increasing threats of excessive heat, drought, flood, storms, and wildfires? There's not an easy answer. The short answer might be fewer people on the earth living more responsibly.

I wonder if home can even be a place when every place is changing and no community or river flows "like it used to." And then I wonder if home might be a state of mind, but that too is problematic because as I learn and grow my mind keeps expanding and rearranging itself as old beliefs prove to be illusions and new questions arise. Perhaps I am asking the wrong questions. Perhaps it's not about finding your home or re-creating what once was, or even simply creating the feeling of home in our minds. Perhaps what we must all endeavor to do is realize that home is more about how we choose to live, and not where we choose to live. I don't know. Do you?

As I looked out of the porthole window of my jet airliner and saw for the first time in almost forty years the mountains I once vowed to never leave, I am reminded of how life always happens to us—without regard to our plans. Resilience and adaptability are the superpowers of house sparrows and geckos. They know that any awning can hold a nest and any building wall will suffice for a day of hunting moths. Songbirds and tiny lizards teach me important lessons about life and living.

Sue Kerver is my friend. In a way, we are battle buddies and have been through a lot together in what seems like a mere flicker of time. We met after she read my first book, *Casting Forward*. She reached out to me and shared that my words spoke to her as she struggled with her own version of post-traumatic stress and the vagaries of life. As fellow military veterans and deep-feeling human beings with unhealed soul wounds, we bonded while casting into a Texas Hill Country stream.

Later, while sitting on the tailgates of our trucks, listening and sharing our mutual quest for "tailgate wisdom," we both chose to make the leap away from comfortable discomfort and toward the yet unknown joys of our newest adventures. If you cannot endure uncertainty, you will never enjoy the certain gifts of adventure.

Sue greeted me at the Bozeman Airport baggage claim, and we smiled and embraced as old friends tend to do. She said nothing about my increasingly aged face, and I genuinely complimented her newly dyed purple hair. Since I last saw her, Sue had made the bold decision to leave her job in South Carolina that was familiar but toxic to both her health and spirit. Shortly after our tailgate talks by that little Texas stream, she accepted a new position in Montana, sold her house, moved sight unseen to the state, put herself through fishing guide school and started her own guide service for women called Two Gals and a Boat, and then met her life partner, Josh Howery, who I'm so happy to share is a great guy. Together they bought a beautiful home near Bozeman where Sue, Josh, her dog Meatball, and his cat Junior now live. That's a lot of change to undertake in two years! There was something else that had radically changed about my friend—she looked supremely happy. I'm so proud of her.

~

In the morning, Sue and I loaded up her SUV and headed out toward Paradise Valley and our destination of DePuy Spring Creek. Josh had to work that day, and while Meatball made it obvious that he desperately wanted to join us on this adventure, it was both inadvisable and non-permissible that we bring him—no matter how charming he might be. So he reluctantly stayed home while giving us the "sad puppy" face as we packed up for the day. In contrast, Junior's squished-up furry kitten face seemed unmoved by our pending departure as we walked out the door—fishing gear in our hands and serenity on our minds.

It's been my experience that while dogs give constant and unconditional love, cats more often than not offer conditional tolerance of human companions that is occasionally offset with moments of inexplicable affection. Thinking about this and about the people I've known in my life, it gives new meaning to "dog person" and "cat person."

DePuy Spring Creek is a lovely spring-fed creek that arises about three miles from its delta, which empties into the mighty Yellowstone River. It runs between the breathtakingly beautiful Absaroka and Gallatin Mountains and is known for its wild populations of well-educated brown, rainbow, and Yellowstone cutthroat trout. I always have mixed feelings about finding these once primal and distinctly North American places filled with wild but ultimately non-native fish that compete with and often extirpate the natives. It's not the brown trout's fault that it doesn't belong here—it's ours.

The water on the spring creek is often thin, usually clear, and so far, always cold enough to support trout and their food items. Like all relatively healthy spring creeks, DePuy is full of life—just as it should be. There is a healthy and intact riparian and buffer zone of willows, wildflowers, and native grasses that add to the vibrancy of the creek. Hatches of mayflies, midges, and caddis as well as terrestrials make this a classic fly-fishing water and a destination of choice for any angler who wants a challenging experience while being surrounded by an absolutely astounding natural landscape.

The quaint "angler's hut" added a bit of unique western character to a morning on the creek, while the owner's replica antebellum house seemed oddly out of place. As I looked out toward the Yellowstone River, which flows beside the creek, and across the valley toward the snowcapped mountains, I tried to imagine this historical landscape with bison instead of buildings and tepees instead of town houses. Wherever I go I wonder what it was like before us.

When we arrived we met up with Jim Wenger, who once served as a U.S. Navy F-4 Phantom pilot during the Vietnam War. He currently serves as the regional lead for Project Healing Waters' fly-fishing program, and is someone who fishes DePuy often and knows it well. He is a soft-spoken, kind, humble, and generous man who quickly revealed himself to be a wonderful fishing companion. I liked him instantly, and I felt grateful to Sue for deciding to introduce us. I'm fortunate in that way. I've met the best people of my life while standing in a river.

When we arrived at the angler's hut, I was reminded of the scenes in the Joshua Caldwell film *Mending the Line* that were shot here. Looking into the waters of the creek, I recalled those scenes of casting, fishing, and laughter. And I remembered the moments when the characters struggled, but also how the act of fly fishing and connecting with Nature gave them new perspectives and abundant solace. It's been like that for me as well—struggle, growth, new perspectives, and new soulful living.

This deeply meaningful and moving film is dear to my heart. As a former U.S. Marine who has suffered from the effects of post-traumatic stress disorder (PTSD) and who found salvation through my connection to Nature via fly fishing, the story of the characters might as well be my story. I remembered how my eyes became moist when I first heard the character named Lucy read my words. You see, my first book, *Casting Forward*, and some of my words were incorporated into this beautiful movie that follows the healing of two U.S. Marine veterans and one civilian woman, as they endure, learn from, and overcome the negative impacts of their life's trauma. We all carry wounds, but to paraphrase the Sufi mystic Rumi, the wound is the place where the light enters us. We can treat wounds as a weight or a way forward. It's a choice.

We strung up our rods while sipping morning coffee at the wooden picnic bench near the angler's hut. A western tanager sang from the cottonwood tree that shaded us from the bright morning sun. Tree swallows dipped and dove as they seemingly swam through the air in search of midges, mayflies, and caddis. And even as we set up our gear and spoke in hushed morning tones, we could see several nice trout holding in the current of a pool near the angler's hut bridge. Jim asked if we wanted to try for them, but both Sue and I offered them back to him, and we watched as he sent several nearly perfect casts into the current, without effect. It was a tough spot in clear water with bright light, but there would be many more chances before this day became evening, so we moved on.

Jim took the lower pool that was closest to the downstream side of the little bridge while Sue and I made our way through some waist-high grass and wildflowers that were bending and bowing in the soft mountain breezes. The sunlight warmed my aged face while the cold mountain air refreshed and rejuvenated my weary soul. The sounds of the creek as it tumbled over rocks and riffles mingled with morning birdsong as I made my way down the embankment and ever so carefully toward a deep pool with fast water and a foam line slicing through its midsection. It looked fishy.

Looking downstream I could see Sue casting into the next pool, and I could just barely see Jim upstream and on the other side of the tall grasses of a midstream island. Just as I glanced his way, I saw him set the hook on what looked to be a good-size fish. His rod bent, his attention went with it, and I heard him let out a whoop and saw the look of utter bliss spreading across his face, which was the face of a peaceful warrior. I whooped back and hollered, "Jim's getting it done and show'n how it's done!" Sue yelled out, "Yeah Jim!" and we all celebrated for our friend, and that was about the time Jim suddenly vanished from view and things went silent—for a moment.

I think Sue had a clearer view of Jim than I did, and I heard her ask, "Jim, ya all right?" I could see him again then, standing there with a big smile and dripping with spring water. He laughed. "Took a swim and lost the fish," he called back. "I'm fine." We all smiled together as each of us picked up our lines and cast once again. And that's another way that fly fishing and life are exactly the same. Sometimes we fall, but we get back up and cast again—together.

Sue and Jim continued to work the pools they were standing near, but after a number of passes I decided to clamber up the high and steep embankment on the opposite side of the creek and walk farther downstream to check out what I might see from above. Once I reached the top I was suddenly awestruck by the scene that rolled out in front of me. There, not fifty yards away, was the mighty and meaningful waters of the

Yellowstone River as it rushed away from the Yellowstone Caldera and onward toward the Mississippi River and, ultimately, the Gulf of Mexico. That's a long way for a raindrop or snowflake to travel.

Just beyond the river was the Absaroka Mountains, which were blessedly still snowcapped in places and snow speckled in others. A meadow of grasses and wildflowers joined the river to the forest, and the forest joined the meadow to the mountains. Across rivers and landscapes the same truth remains: Transitions are where the action of life plays out.

I am so in love with transitions when they are made from Nature, and so fearful of them when made by my fellow *Homo sapiens*. The first is transition made of coexistence and balance between what's offered and what's given. The second is most often the result of taking without reciprocation, of extraction without empathy. It's the difference between romance and rape. I am forever a hopeless romantic and wistful dreamer. I remain hopeful that we will evolve into a species that values love over lust.

As I walked along the high creek bank, I looked down into the water and up into the sky while keeping a wary eye on the willows for angry moose and grumpy bears. I could see trout rising, although I couldn't see anything hatching on the water. I walked through the grassy embankment watching for grasshoppers and other terrestrials. Finding none, I stuck with the nymph rig that Jim suggested.

With climate change, I have found the same situation everywhere I've traveled. Either the area is sunbaked, drying up, with fish sweltering and whole sections of river dry to the stone, or the rivers are higher, faster, and colder than usual with excessive rain and flooding being the new normal. Montana was the latter. So as much as I wanted to tie on a dry fly, I put my faith in the deep-drifting nymph rig. I slowly, carefully, mindfully worked my way over and into position for a cast toward some big trout that were holding and rising in a long pool of quick water.

Once in position I noticed another nice-size fish swimming and swirling and taking something just below the surface in a tiny pool closer to the bank, and closer to me, so I cast to that fish first. It was a tricky presentation for a clunky nymph rig. I had to drop the rig just upstream and into a crease of current so that I wouldn't drop my line over the fish but instead the current would drift the nymph into the trout's little home pool. I managed it several times (without result) before the gig was up and mister fishy began to sulk.

So I moved with as much stealth as my aging legs allowed over the slippery stones and through the quick waters of this sacred spring creek. Red-winged blackbirds sang from cottonwood trees and fence posts. Sunlight filtered through the willows, and a few darkly colored caddis lifted up on fresh wings into their new beginning and ultimate ending. If only my cast was as magical as that moment, but it wasn't, and after a few more tries I knew that these fish would remain safe and sound in the here and now. It mattered not. This was my version of heaven, and I found myself wishing I could wish away the artificial antebellum mansion and replace it with the cabin that my once-young wife dreamed of with her once-young husband. I heard myself repeating the words I spoke so long ago: "I never want to leave."

There's a touching and poignant scene in the film *Mending the Line* in which the four main characters—Ike (Brian Cox), Coulter (Sinqua Walls), Lucy (Perry Mattfeld), and Harrison (Wes Studi)—share a streamside lunch together in the same spot where my friends and I now ate, conversed, listened, and laughed. I have come to believe that the story of our lives is the ultimate work of art we might ever create—if we are willing to paint with purpose and passion. Art imitates life and life imitates art. The two are inseparable as the best aspects of human nature. And here we were doing in real life what screenwriter Stephen Camelio wrote and director Joshua Caldwell interpreted and reimagined so beautifully. We were living out what the actors portrayed. I suspect the biggest difference is that once either the director or the weather calls "cut" on the scene, the actors will move on to another job, but we retain this memory and our friendship beyond this single moment.

We were playing together, not playacting. The friendship and fellowship were as real for us as were our words and wounds. And for all our battle scars, we three peaceful warriors sat together beneath the cottonwoods and beside the rushing waters—exuding gratitude and joy. We managed this level of living perfection in a single take.

After lunch Jim invited me to go with him streamside to do a bit of scientific inquiry. We were going bug hunting, one of my favorite childhood pastimes. Stepping into the creek, Jim stretched out a screen as I shuffled my wading-booted feet and overturned submerged stones. After a few moments of doing the Texas two-step while Jim acted as dream catcher, we lifted the screen to see what we had captured. The catch included earthworms, aquatic worms, a few small mayfly nymphs, a couple caddis nymphs, and one smallish stonefly nymph. In hindsight, I wish I had followed my inclination earlier and switched to a hopper-dropper rig with a caddis dry fly and a stonefly or mayfly nymph dropper. But instead, I stuck with the nymph rig because that's what Jim was using, and he had quickly become a new hero for me.

Here was a man who served his country with courage in a war we probably should have never entered. (Most wars fit this category for me.) He came home to live a good life and one of service to his community and the veterans who came after him. And now here we were together finding solace and sense in the waters of a spring creek. Life is truly grand if we choose to live it wisely.

Sue decided to work the lower pool not far from the angler's hut while Jim and I walked upstream along a trail that was surrounded by soft, high grasses and willow thickets. I worked my way down a steep embankment and crossed the creek so that I could blind cast into a likely pool while Jim kept walking until he vanished around a distant bend. I was alone on the river with only the wind in the willows and the soft hush of the water to keep me company. It was peaceful. It was perfect. Life was good indeed.

The fishing was wonderful; the catching was elusive. The morning had unfolded with Jim catching a single fish, falling in the water, and hanging half of his clothing in the sunshine from a nail on the angler's hut wall. Neither Sue nor I were able to elicit a single strike no matter how diligent we were at reading water, finding trout, and presenting our offerings to them. Still, we remained happy and hopeful, with each new drift holding promise and each bend in the stream offering new questions to be answered. Angling is the practice of asking and seeking the answers to some of life's best questions—the questions about the true nature of life and living. Trout and mayflies are wise teachers for a novice like me. I retain my beginner's mind no matter how many casts I send forward or retrieves I make.

The sun and its warmth began to vanish with the rising winds that crossed the meadows and the falling coal-colored clouds that surrounded the mountaintops. I'd walked, waded, and wandered upriver without result, but nothing was lacking. The stream banks swayed in grasses of green and gold and wildflowers of yellow and blue. The swallows sailed overhead as the creek's clear, cold waters wound below me. I felt peaceful, as in full of peace.

As I worked my way from pool to riffle and onward toward our rendezvous site at the dirt road crossing, I heard the sound of Jim whooping in celebration of another fish caught, and I cheered him on as he raised it briefly and then set it free. I was happy for him, and we both smiled as we met up at the riverbank trail. About that time Sue came around the corner and shared that she'd had as much luck at catching a fish as I had. We started to walk toward the dirt road along a narrow, muddy trail that filled my mind with images of moose and bears looming and lurking just beyond the edge of the willows.

Just as I was thinking that our day was nearing its end, we came to a narrow game trail and Jim said, "Let's try one more spot." As he led us down the twisted path and back toward the creek, I must admit that I wasn't very hopeful. Lucky for me, Jim kept the faith.

He pointed to a stretch of water with bright green aquatic vegetation swaying along one side of a deep pool and a deep grassy undercut on the other side. "That's good in there," he said, as he turned and began casting into a seemingly lesser pool just upstream. Sue decided to walk down to the next bridge to fish the deep pools above and below it, and we said we'd be along to meet her in short order.

I waded ever so carefully and cryptically toward the pool, and for a good long while I just stood there watching and waiting to discover what this spot in the river was trying to teach me. I watched the way the water moved and opened my mind and soul to what it might mean, and in my mind's eye I submerged myself in the currents and looked up from the undercuts ever watchful for ospreys and kingfishers.

The rain began to fall softly, and I looked up into the now slate-colored sky and back at the half-submerged cotton-topped mountains. That's one of the lessons: Life is change, so we adapt. So I cast as if I were the nymph in search of a trout. I heard the words in my head: "Think like a trout, act like a nymph." And that's when the indicator submerged and my rod tip rose and then bent under the gravity of the rainbow at the other end. At last, I had managed to connect to a DePuy Spring Creek rainbow trout in Montana's Paradise Valley. Happily, I raised the fish in my wet hands and Jim cheered. As I watched the lovely little rainbow swim back to the pool where I found her, I couldn't stop smiling. I was in heaven.

When we reached the bridge where the dirt road crossed the creek, Jim said he'd walk up to get his truck and drive us back to the hut while we fished the bridge pools. Sue was dutifully working each bit of promising water, but I decided to snip off my fly, reel in, and sit on a bench beside the tumbling water just as the rain began to fall, this time in more frequent drops of greater size. At first I looked out to the wall of woolly clouds that now obscured the vast Absaroka and Gallatin Mountains while allowing the raindrops to roll off the brim of my hat, feeling deeply peaceful and content. But then I raised my face into the rain and allowed it to tumble from my open eyes like it did so long ago—memories of just holding on.

There was a time that feels like a lifetime ago—as if it were another life I once lived and from which I was resurrected. In that long-ago time I found myself sitting alone on an old wooden bench that was embossed with a brass memorial plaque with the name of a dead man who once sat where I was sitting—another life seeking the same solace. And so it was back in that long-ago time that I found myself sitting beside Cibolo Creek with the rain falling softly between the cypress trees and into the creek's winding currents, and I wondered if I wanted to keep breathing.

This was a time when my PTSD had been most aggressive in choking the life out of me and turning all of life's color sepia-tone. As I sat there remembering that long-ago time of despair, I felt overcome with gratitude that I had made the right choice back then. As I watched my dear friend Sue casting her line and my new friend Jim walking down the long dirt road in the gentle rainfall, I felt so grateful that my love of family and my connection to Nature through fly fishing caused me to continue to breathe. An intimate connection to Nature is good medicine. It gives us a healthier perspective.

Life and fly fishing are metaphors for each other. And this day of many casts and one final connection was the gift of Zen-like understanding. You see my friends, this day reflected the lesson I want every half-broken human heart to learn and remember. No matter what life may give you or seem to take away, there is always hope. When the rain falls, embrace it. Let those life-giving droplets cleanse your wounds and grow new wildflowers in your soul. Embrace struggle for the gift it truly is and accept loss for the precious lessons it offers to teach us. It is in hardship that we are all defined. Expectations are the root of all suffering. Fish or no fish, every day on the water is a gift.

That evening, Sue, Josh, and I attended a barbecue they were hosting on my behalf at the beautiful home of a friend. The house was high up on the foothills of the mountains overlooking Bozeman on a large, open hillside surrounded by meadows filled with grazing horses and mountains filled with possibilities. I liked seeing the horses but imagined them as bison. I loved seeing the mountains and wondered what other living beings were looking down toward me as I looked up toward them.

A dog came around my legs to greet me. I smiled and spoke to him in soft tones and thought of the "service dogs" that help people with issues like anxiety and clinical depression. I've been there. Sometimes, I'm still there. The road from suffering to thriving is a long one but well worth the journey.

Josh was getting the fire going on the grill as people were beginning to arrive—to meet me. He and Sue had been so kind in accommodating my efforts to eat healthier after my heart disease diagnosis, and he was grilling some veggie burgers for me next to the beef burgers everyone else was craving. I've discovered that how we treat our bodies and how we treat our friends, families, communities, and planet are quite the same. Healthier choices lead to healthier outcomes. Still, those roasting burgers looked and smelled amazing! Small steps forward.

It seemed wonderfully absurd that anyone wanted to meet me and absolutely wonderful that I was about to meet them. These were the people of Sue's circle. There were fly anglers and other outdoor enthusiasts, and people from Sue and Josh's church, and people like Jim—veterans who donate their time and skills to help others discover the healing powers of fishing, friendship, and fellowship in Nature. Programs such as Project Healing Waters, Reel Recovery, Casting for Recovery, Warriors & Quiet Waters, and the Mayfly Project all work to help people connect with Nature and the best of human nature as a way to healing and health. I am a strong advocate for such selfless organizations. We can all find ways to help each other and this beautiful blue planet that is our one true home.

I loved every moment of this day and evening. As the sun grew closer to the edges of the mountaintops and the air began to chill, we sat by the fire speaking in hushed tones and laughing with vigor. We shared conversations in-the-round and more than a few that were one-on-one. And then I found myself walking away from the warmth of the circle and into the chilling breezes that were rolling in from the distant Absaroka and Gallatin Mountains. I found my imagination traveling toward the spring creek and wondering if I might ever stand in its healing waters again. It mattered not. I have stood there and have received that good medicine. And in that moment, as the chilled air reminded me of what a blessing it is to simply keep breathing, I was grateful. Nature heals us. It's past the time to return the favor.

Chapter Eight

Floating the Big Hole, Madison, and Missouri Rivers, Montana

Humankind has not woven the web of life. We are but one thread within it. Whatever we do to the web, we do to ourselves. All things are bound together. All things connect.

—Chief Seattle

When we arrived at the Big Hole River, it was peaceful. There was a soft mist falling from the ash-colored sky, and the morning birds were just beginning to sing the good news to each other that they had made it through the long, cold night. I know that song, even if I often have no one to sing it to but the morning birds and memories of loved ones who have crossed over. It's comforting yet foreboding. It reminds the singer that waking is not guaranteed and every day of life is a beautiful gift. At home in Texas I often hear the wrens calling out at dawn, "I'm here! I'm here! I'm here!" I always wish that I had a way to call back to them, "I'm here too!"

In a way, we all sing our morning songs in the hope that someone sings back. And when they do, we are comforted and reminded that we are not alone in our mortal aloneness. Whenever I hear the birds singing as daylight dawns, I feel as if they are singing to me. I am fully aware that fish are not rising to see me, but still, I do not want to live in a world devoid of singing birds or rising fish. As we strung up our rods,

I became thoughtful. I did my best to remain in the moment, but I felt the pull of history dragging me back to a time when this river was not so peaceful.

When I first walked up to the river, it was running high, fast, cold, and wild. It almost felt menacing. Meatball was eager to take his place in the stern of the craft where he could keep an eye on Sue and absolutely everything else that might be going on around him. He looked so ready for adventure with his bright yellow doggie flotation vest and wagging tail.

I could hear the murmur of Josh and Sue speaking of the coming day as they created their leaders and tied on their chosen flies. This was their home water, so I imitated their choices in the hope that their intimate knowledge might allow me to know this river and its fish better during the brief time I had in these hallowed waters. I needed to experience the Big Hole River just as it was and not as I had imagined it. Warm introductions and sunlight always help break the ice.

Josh rowed us out into the quick current, and we immediately became connected to the river, and she connected to us. We rolled over her waves and into her vortexes and across her softer runs. Sue and I wasted no time and began casting toward the edges and eddies among the half-submerged willows. We were tossing "hopper-dropper"–type rigs, which I prefer to the usual nymph and bobber contraption that may be effective in catching fish, but to me feels as elegant as swinging a medieval mace. It might as well be a worm at the end of my childhood cane pole, complete with an oversized plastic bobber—not that there's anything wrong with that. It's effective, but clunky.

Much like my experience in the Wood River drainage of Alaska, the accelerated global climate change that is drying up the rivers in Texas is also filling up the rivers in Montana in an equally destructive way. Fish and trees were being buried under too much water after unseasonably heavy rains and runoff. Hatches were as turned off as the rivers. The new normal is that nothing is normal anymore.

In between sections of castable water, we came to stretches of choppy, whitecapped, rock-strewn river where Sue and I would sit and Josh would row while Meatball shifted from side to side in excitement, innocently making things more difficult for our oarsman. But while Josh worked to keep us upright and moving in the right direction, I used these moments to pick my head up from the business of casting and pay attention to the vast landscape that unfolded all around us.

Montana is a land of many landscapes. Each river seems to have its own homeland, and no two watersheds are quite the same. Along the first section of the Big Hole, I looked up into the naked rock pinnacles of the cliff faces that were populated by conifers about midway down, then with cottonwoods along the river and a wall of half-submerged willows at the water's edge. With the river running so incredibly high for this time of year, many of the willows were completely submerged and therefore a snag hazard to be reckoned with as we simultaneously tried to present our flies close to the bank, but not so close that they became a permanent part of the ecosystem.

After the first few turns in the river I got a strike and briefly hooked a fish, but I must have done a poor job in my hookset because it made a single flip and headshake before it was free again. There was nothing to do but keep casting, and we did. We cast again and again as the river pulled us ever forward. And that's the thing about floating on quick water—like life, we often get only one brief chance to connect, and then the river and life moves us out of "casting range." Life and fly fishing are much the same. When it comes to our relationship with each other, our tribe, or our planet, often there are no second chances.

Josh lined us up with a nice stretch of riverbank that consisted of overarching cottonwoods, overhanging willows, and an assortment of underwater deadwood. The pace of the current here was optimal in that it was just enough to keep us moving but not so much to keep us from having some nice long drifts over fishy-looking water. I was the first to hook up and was happy to get the skunk off me as I brought a smallish but beautifully colored rainbow trout to the net and released it, seemingly unharmed. Sue connected with and landed a much bigger and just as beautifully decorated rainbow. There were high fives all around as we regained our forward momentum and began casting, drifting, and recasting in a Zen-like meditative state. For all the tumult of the overflowing river, my spirit felt calm, mindfully in the moment and soulfully at peace.

I picked up another fish just before we decided to anchor along a shoreline sheltered by a grove of cottonwoods that were alive with birds of many shapes, sizes, and sounds. A pileated woodpecker was calling insistently from deep inside the cottonwood trees. Halfway up one old tree I could see a nesting hole, but I never got a glimpse of the prehistoric-looking bird with its bright red crest and massively powerful beak. I did get a glimpse of several ravens that were calling from the top of some distant Douglas fir trees. In the meadow on the other side of the river I could see and hear a mixture of red-winged and yellow-headed blackbirds clinging and calling from the low willows and tall grasses. A loon called in the middle distance, and a pair of Canada geese paddled downcurrent with goslings in tow. Every so often, a western tanager would flit between two trees with its bright yellow body; black wings, tail, and back; and striking red head on the breeding males. No such magic exists in a manufactured world. This was real. This was special.

As I ate my almond butter and fruit spread sandwich and drank my second cup of coffee, I found myself enjoying the serenity of the moment and the sheer luxury of warm coffee served outdoors from a metal camp cup. Each day, Sue brought us a thermos of dark, flavorful coffee and two "Rep Your Waters" metal cups. We each had one: Her cup was orange red and mine a darker blue green, with a vermiculite trout pattern on each. From back in our Texas "tailgate wisdom" days, we had made it a tradition to tap our coffee mugs together as a toast during a morning fishing break. Coffee is my elixir of life. I am enthralled by its aroma, flavor, warmth, and the comforting, almost religious practice in its preparation, presentation, and enjoyment. Whenever I travel to places that prefer tea, I bring my own coffee.

The morning began with an autumnal coolness to the air that felt like the end of something and the beginning of something else. The charcoal-colored clouds allowed only the mere suggestion of sunlight. Even with the falling temperatures, nothing else was falling. There were no golden leaves or increasingly shallow waters. There was no doubt that this was a summer's day devoid of warmth or light. Everything around us was dark, deep, and daunting.

The river rolled with big voluptuous swells followed by smaller knife-edged waves that chopped and slashed at our hull. Conflicting currents competed on their inevitable journey away from the mountains and toward the sea. People and rivers can be like that—pushing away pieces of themselves as they simultaneously yearn for connection to the whole.

In time the clouds drifted off as they tend to do, and the sun began to shine as it always does—eventually. We continued our journey down the Big Hole with Josh expertly moving us from one hopeful bankside to the next hopeful bankside. Sue and I kept casting, unperturbed by our dismal catch thus far. As for Meatball, he was asleep in the stern of the boat with his head resting on his stuffed Tigger toy that gave him comfort whenever things seemed uncertain—as they always are.

We kept casting, but the river was high and cold, and the fish were not having any of it. The salmonfly hatch never came, and the reports were that the Big Hole River was suffering a massive trout population crash for reasons yet unknown. I'm not sure what is causing the wild fish population on the Big Hole to suffer, but I have a healthy suspicion that the activities of *Homo sapiens* are involved.

An eagle flew overhead, also looking for fish and finding none. As we rounded the last corner to slide under an old, rusted bridge, I was thinking about how peaceful this float felt, and how calming the water, breeze, birdsong, and sunlight on the mountains made the entire experience. I looked across a meadow that held a mix of tanagers, blackbirds, and sparrows, each singing their springtime songs in the autumnal cold.

Peaceful. I wondered if it felt that way on the morning in 1877 when Colonel John Gibbons's U.S. Army troopers opened fire on the sleeping Nez Perce village, killing men, women, and children and initiating the two-day Battle of the Big Hole. Shortly after the Union Army had fought a war that ultimately ensured some level of freedom for African Americans, the same army turned its guns toward the task of subjugating Native Americans across the West. These forces were ultimately led by none other than the famed Union generals Sheridan and Sherman, both noted for their use of "scorched earth" warfare tactics that today might be described as war crimes. Among the dead were mostly women and children on the Nez Perce side of the conflict, with Chief Looking Glass declaring that they also lost some of their best and bravest fighters. Colonel Gibbons's troops suffered the loss of about 30 percent of his entire force.

The Nez Perce fought to protect their families, having already lost their ancestral homelands to gold mining and repeated treaty violations by the US government and Euro-American settlers. From the place that the Nez Perce once called home, they were being hunted and exterminated as they fled east across the treacherous Lolo Pass and down the Bitterroot Valley. All they were trying to do was find a new place to call home. All they were trying to do was make it to the relative safety of Canada, where they might rebuild their lives in exile. They were never given that chance.

I still dream of a world with more meadows than minefields. Meadows are the creation of Mother Nature; minefields are the manifestation of our collective choices. It simply depends on which world we wish to cultivate. I prefer wildflowers over barbed wire and wild rivers over canals. Don't you?

The vast green valley that cradles the upper Madison River between snowcapped mountain ranges is quite simply one of the most beautiful places I've ever known, and I've known a lot of beautiful places. We entered the Madison River watershed through the small town of Ennis—a town that seems to thrive on fly lines, leaders, and drift boats. A bronze sculpture of an angler stands in the middle of town, complete with fly line and leaping trout. I could see multiple fly shops and outdoor recreation stores, one of which we walked into in our effort to get the latest news on the condition of the river and the fishing.

We did this everywhere we fished. When we walked into the shop I noticed that just inside the front door was a large stuffed beaver holding a PBR beer can and flashing its bucked teeth in our general direction. When I was a young man, PBR was a blue-collar workingman's beer, but apparently now it's become the trendy brew of choice for many of the young guides. Who knew?

Josh and Sue did the recon and talked with one of the shop workers about flies and rigging and river conditions, while I acted more like a tourist looking around at hats and coffee mugs—although I did pick up a few leaders. After an extensive conversation about how the Madison was running "unusually high, fast, and cold for this time of year," Josh and Sue invested in a few select flies that had been suggested by the young man working in the shop as "flies that are working on the river . . . now." I was skeptical because we had done the same thing on the Big Hole and DePuy Spring Creek without any noticeable result. We might as well be casting bits of Sue's purple hair tied to a hook. And I expressed my skepticism to my friends when I said, "I'm beginning to believe that the 'flies that work' are whatever is in inventory that needs to move off the shelf." We all laughed that nervous laugh that people have when a joke isn't really a joke.

When we got to the river it was rolling high, deep, fast, and cold, but the sun was shining with just a few soft white clouds drifting over the snowcapped mountains. Due to the unusual heavy rains, the meadows were as green as Ireland, and bright yellow wildflowers covered every edge of the river. Flocks of black, yellow, and red western tanagers flitted from willow to windblown bunchgrass, and a bald eagle flew downstream on outspread wings that flapped not even once. Eagles possess such mastery of the air and connectivity to the water that I envy them. Still, she looked hungry. I wondered what that meant.

Unlike our brooding and windswept day on the Big Hole, it was temperate this morning. But just beyond the bright blue morning skies was the dark gray foreboding of the afternoon. We launched and almost immediately the drift boat was sent like a bullet into the twisting currents as Josh worked hard to manage us around the many massive, rounded rocks that waited just above or below the churning surface.

Once again Sue insisted that I take the bow of the boat since I was their "guest" and they could "fish here anytime they wanted." I did not resist her logic or generosity. River conditions everywhere were less than ideal, to say the least, and I knew that I'd need whatever edge I could manage if I wanted to connect with some Madison River trout. Toward each bank we cast our fly shop–recommended rigs of massive silver-blue streamers with tiny nymph droppers. I love streamer fishing. It feels like poetic warfare—both intuitive and interpretive while also being quite aggressive.

According to the young guy at the fly shop, "big rainbows" were sometimes hitting the big streamers and other times being "attracted" by the streamer but striking the nymph. I'm not sure if this was true or just another potentially comforting story that is told when it's uncomfortable to say, "The fishing is super slow on this high water; I have no idea what might work." But no matter if a fictitious story or factual sharing of local knowledge, after a few minutes of casting and stripping in line and recasting in rapid-fire succession, I got a strike from a good-size fish that shook its head and made a single leap into the air before tossing the tiny nymph back into my windblown face. I think it was a whitefish. I guess I'll never know—for sure.

When things like this happen on a day that you know might be a tough day of fishing, it gives you that curious mix of feelings: regret, self-recrimination, and a dash of hope that maybe things won't be so tough after all. The regret is that you failed to land the fish and discover its true identity. The recrimination is for your inadequate hookset or failure to fight the fish well. And the hope is that this is just one of many opportunities to connect with something wild and free in a beautiful setting with wonderful friends.

But that's not the way things unfolded. The low-hung clouds began to roll over the mountains and across the once blue skies. Rain began to pelt us, softly at first and then with more vigor and urgency. The warm early summer temperatures began to plummet, and an early winter mix of ice-cold rain, hail, and sleet fell. I watched for snowflakes to fall, and seeing none I kept casting.

In time, we three intrepid adventurers put on every layer of insulating and rainproof clothing we had, all the while poor Josh labored dutifully at the sticks. Meatball shifted uncomfortably in the stern, again making Josh's task all the more difficult. Sue did her best to help her noble pup and equally noble partner, but even the best of boat puppies isn't going to be comfy in a rocking boat being pelted with ice. And through it all, Josh kept rowing without respite, and we kept casting without catching, and we all continued to share an amazing adventure together. I was supremely happy. No amount of cold rain was going to dampen my grateful spirit.

I remind myself that life unfolds as it will. Each cast has its own character, like fingerprints we send out into the wind and water. What we make of life is our choice. My father used to say that "into every life, a little rain must fall." But I never forget that we need the rain as much as the rainbows. We need the texture that struggle gives us, and the chance to learn who we really are and what we're made of. There's a good reason why every cast ends up looking like a question mark unfolding into an exclamation point! We ask the question so as to discover the answer. If we knew everything, we'd have nowhere to grow and no reason to go on. What's the point? I need mystery in order to have a reason to seek mastery.

There were other drift boats floating down the river that day. Each one seemed to be a local guide with one or two clients onboard. I have the sense that I can tell the difference between a guide and client boat and one like ours, which just had three friends floating together. I am hopeful that Meatball considers me a friend, and if he does, that makes four friends floating together.

Only once during the day of drifting did I see anyone's fly rod bending besides my single brief hookup to that suspected whitefish. It was a sport in another boat who had hooked a willow tree, and the guide had a disgusted, "let's call it a day" look on his face. I felt a bit sorry for them both, because they could have just as easily been having the same superlative day we were having in our boat. Boats are like islands and lifetimes; they contain whatever you bring to them.

During the final mile or so of the float, I did what I often do and decided to snip off the fly, reel in the line, and sit peacefully in the bow with the cold air bracing against my battered old face and the reflection of cloud-crested snowy mountains bouncing off my grateful eyes. Placid herds of black cows dotted peaceful meadows of green grass and yellow wildflowers. Two bald eagles sat in a towering old cottonwood tree near shore—one sat on the trees' windblown crest while the other sat on their massive nest that seemed too heavy for even the stoutest of trees to hold.

But the tree held the nest and the nest held the future of her species just below the fluff of her feathers.

I thought of the first time I read Rachel Carson's *Silent Spring* and how close we came to extinguishing these birds forever. I don't want to live in a world without eagles. I really don't. What's the point of living my feral lifetime in a world devoid of wildlife? It's not for me.

For me, life must be wild. I wish we humans would learn that we cannot own each other anymore than we own dogs, cats, canaries, or the land itself. We own nothing. We simply live and die, together.

At the boat ramp I saw the guide parting ways with the client who caught the willow tree, and afterward he looked over at us and said, "It was rough out there!" He looked unhappy. I felt sad for him. You see my friends, just like the two men in the other boat, we had traveled many miles down the same wild river in wind, rain, sleet, and hail. And just like the guide, Josh had rowed the entire eight hours of the float without rest or respite—although both Sue and I had offered to help. I think Josh saw this as a challenge to be met and a joy to be accepted. And just like them we didn't land a single fish even after hours of casting and retrieving while riding the waves and chop of the windblown river.

But as I watched him and his client walk away with their heads bent downward toward the earth as if asking the planet, "Where were the fish?" I realized that we had little in common. Two boats traveled the same river, but the two tribes of travelers experienced different days.

I enjoyed every cast. I loved every icy moment. I felt grateful for the way the cold air caused the warm coffee to feel like magic. I felt joyful for the greenness of the grasses and the grayness of the skies. And my head was up, even in the rain—facing into the bracing wind. Life was beautiful. Some thoughts are worthy of repetition, and this is one: "Everything Is Perspective; Perspective Is Everything."

The Mighty Missouri was running at about 16,000 cfs a few days earlier but had calmed to a still rambunctious 13,000 cfs by the time we arrived in Craig, which seemed more like an accumulation of fly-fishing shops, restaurants, and gas stations surrounding a boat ramp than anything else. I liked it, and after a stop at one of the local shops for more "what's working" advice, we launched the drift boat and began the longest float trip we'd tackled to date. I could see Josh's weariness from the past few days, but he still insisted that he'd be fine, and so Sue and I were once again allowed the pleasure of simply fishing and enjoying the ride. Not a bad life if you can live it. Not bad at all.

The one saving grace for Josh was that Sue had agreed that with me in the bow and her sharing the stern with Meatball, and Josh having to contend with the extra shifting weight in the already shifting currents, it was best for all involved that our favorite pup stayed home with Junior the fluffy cat. I'm relatively sure Meatball didn't agree with this decision, and we were treated to more "sad puppy" face as we packed up that morning and left the house without him. But soon enough I would be back home in Texas and Sue would be back up in the front of the boat while Meatball and his stuffed Tigger would have the stern all to themselves—as it should be.

The Missouri was indeed wide, deep, vast, and mighty, but when we pushed off and into its currents it also felt tranquil. And if the Big Hole felt like late autumn and the Madison felt like early winter, the Missouri was more like springtime trending toward summer. The sky was blue and cloudless, and the sun was high and unobstructed by mountains. Everywhere I looked along the shoreline the grasses bent and wildflowers burst into view in every imaginable shape and color. White pelicans flew overhead looking prehistoric in a state where fossils and artifacts litter the landscape and human history echoes around every bend of every river.

As we drifted downriver, I thought of the history of this place. I thought of what it might have been like before the first humans migrated here over ten thousand years ago. I thought of the Clovis point hunters that extirpated much of the megafauna that once made North America more vibrant than Africa's Serengeti Plain. And then I imagined the various indigenous tribal people who lived in and around these waters before and after the arrival of French fur traders who traveled down from what would become modern-day Canada. Finally, my mind drifted to that long-ago nautical journey of a band of U.S. Army explorers who had been dispatched by the president of the United States to explore the newly acquired "Louisiana Purchase" and find a river passage from St. Louis, Missouri, to the North American Pacific Coast.

The Lewis & Clark Expedition, also known as the "Corps of Discovery," launched in the spring of 1804 from Camp Dubois in Illinois, heading toward its eventual arrival at Fort Clatsop, Oregon. The expedition was commissioned by then President Thomas Jefferson shortly after the completion of the Louisiana Purchase in 1803 with the goals of exploring the newly acquired territory, finding a practical and commercially viable waterborne route to the Pacific Coast, and establishing a "first discovery" claim on the region before competing European nations might do so. The band of approximately thirty soldiers and civilians were also charged with cataloging the flora and fauna they encountered along the way, with the dual objectives of increasing scientific knowledge and solidifying the legal claim of the United States over this land and its indigenous people.

And this leads us to the more insidious part of the expedition's goals—namely, to establish trade and cooperation with the various Native American tribes that lived along the Missouri River and its tributaries, and if not cooperation, then sovereignty and subjugation by threat of force. To this end, the expedition was well supplied with various trinkets for trade including newly minted "Indian Peace Medals" with the likeness of Thomas Jefferson on the face of each medal. These items were intended to help facilitate trade and cooperation. The expedition was also well supplied with modern weapons, which were on full display, a less than subtle attempt to imply the new reality of US sovereignty and the inevitability of submission.

As we know from how history played out, the suppression of indigenous cultures and identities became another stain on our bonny tricolored banner. Ironically, if it were not for the warm welcome of most of the Native tribes for the soldiers and civilians of the Corps of Discovery, the expedition would most likely have ended in starvation and failure. And now as we poison our air, land, and waters and feel the effects of our actions beating down on us in the form of excessive heat, floods, storms, and the melting of our polar ice caps, it seems that once again we have lessons to be learned from those humans who lived here first.

We started out casting a dry fly and nymph dropper rig that was set up to run deep due to the high water in the river. As we slipped under the first bridge, Sue almost immediately hooked and landed a good-size whitefish that jumped several times and gave her a solid fight all the way to the net. As much as I know that for some reason whitefish are treated as a "lesser fish" by so many anglers, I don't see them as such, and was hoping to catch one myself. And they seemed to be rising to the surface everywhere, smacking the small, pale mayflies that we were seeing drifting on the water in mini flotillas.

We were all feeling optimistic after Sue's initial catch and release on the dry-and-dropper rig. We were using a parachute Purple Haze dry fly and matching nymph dropper as suggested by the local fly shop in Craig. Fish were rising all around us, and it seemed that all of them were whitefish. Sue and I kept casting and drifting with as much skill as we could muster, and Josh kept us in the zone among the rising fish, but nothing we tried could induce a strike on our imitations. The flies we were tossing looked a lot like the ones I was scooping up in my hand. It was a real head-scratcher, but trying to figure out the puzzle is all part of the process.

One of the unique attributes of the Missouri as compared to the other Montana rivers I have explored is its profusion of massive eddies and whirlpools. And we invested a fair amount of time and energy fishing them, as fish were rising along the edges and picking off the same mayflies we had seen farther upriver. Josh expended no small amount of energy giving us multiple shots at fish along the edges and even inside the swilling waters.

Fish were rising everywhere, but inexplicably ignored everything we tossed at them. And it continued like that as we paddled past the feeding mule deer and the islands covered in white pelicans, and the cliff face where three bighorn ewes and a lamb were scrambling and nibbling bits of greenery that managed to survive in the cracks between the stones. I've always loved those tiny plants that manage to survive against all odds. It seemed a shame to see them eaten, but that's the nature of Nature.

It seems surreal looking back at that day of sunshine and wildflowers and more fish rising than I could ever count, and yet catching only one fish in fourteen miles of fishing. And we didn't actually see any trout—only whitefish. If you think about it, and you should, almost none of the fish we *were* trying to catch were in these waters when the Corps of Discovery floated this river. Back then the brown trout were still in Germany and Scotland and the rainbows were still in Alaska—where they belonged. But as we moved in and moved the indigenous people out, we brought with us the trappings of the places we were fleeing from, including our illusions.

It's not like things were all perfectly peaceful here before we killed off the bison and replaced them with cows. For all their native wisdom, the indigenous people had their own battlefields where they flecked wildflowers with blood. And the soldiers of the U.S. Army may not have fared so well if they hadn't found an ample supply of Native American scouts who were willing to help them hunt down "the other" tribes.

We're all human. We're all living within the limitations of our opaque understanding, just one step out of the cave even as we launch into space. Human certainty about anything is hilarious. Almost everything we "know" is in actuality faith and fiction.

It really doesn't matter if we're in a drift boat on a river, on an island in the sea, crossing some distant desert, or inside the confines of a city. Whatever we bring there is what we find there. Any place can be paradise or hell—it's our choice, and our choices have consequences. I wish with all my broken and battered heart that we'd make better choices—now. It's almost too late. I don't want to fish in empty rivers. Do you?

Part III

Grand Teton and Yellowstone Country, Wyoming

Chapter Nine

Wading the Firehole River, Yellowstone National Park, Wyoming

Be a lamp, or a lifeboat, or a ladder. Help someone's soul heal. Walk out of your house like a shepherd.

—Rumi

Beneath the mountains, meadows, and meandering bison of Yellowstone country lies one of the largest volcanoes in the world, and there it slumbers while awaiting some uncertain moment when it might once again rise up, and in that instant turn Paradise Valley and much of our beautiful blue planet into a "Paradise Lost." It's not certain that this supervolcano will ever erupt, but it somehow seems unlikely that it won't.

The Yellowstone Caldera is massive. The first known eruption, which took place over two million years ago, is estimated to have dropped more than six hundred cubic miles of ash on the surrounding landscape and into the atmosphere while creating almost unimaginable destruction. The last eruption, slightly over 600,000 years ago, left a crater roughly fifty miles in circumference at its widest point. What if even a relatively small eruption occurred now? The difference from then to now is that now humans are here. Even a relatively modest eruption of the Yellowstone supervolcano would certainly be the most devastating natural disaster in human history. Knowing this reality caused me not an ounce of fear as we crossed into Yellowstone country; I am much more fearful of the unnatural disaster that I see looming in the much nearer future. The most tragic catastrophe is one that could have and should have been avoided.

The morning was just cool enough to be pleasant and clear enough to make the drive from Paradise Valley to Yellowstone National Park simply spectacular. Josh was exhausted from three long days of rowing, so he slept in the back of Sue's SUV while we sipped coffee and shared meaningful conversation up front. Sue is the one who coined the phrase "tailgate wisdom" for these conversations we share about life and choices and how our choices act to paint the portrait that becomes our life's story. It's important to edit often and well. Once it's "gone to press," there is no turning back of pages or history: *Iacta alea est* ("the die is cast").

Before entering the park, I picked up a coffee at a stand in West Yellowstone just before a quick visit to Big Sky Anglers for the usual tips on "what's working" on the river of the day. In this case our river choice was the Firehole, which as its name implies flows in the center of the Yellowstone geothermal zone, and as such is surrounded by the planet's largest concentration of active geysers and hot springs. It is also an area of snowcapped mountains, waterfalls, wilderness, wildlife, and waterways that support an entire ecosystem, which happily includes wild trout in abundance.

I choose the term "wild trout" because the Firehole, like so much of the modern American West, is populated with non-native trout that have the potential to outcompete or cross-spawn with the native cutthroat. I often hear local anglers across America misusing the term "native" by referring to non-native trout that have adapted to and begun spawning in the waters where humans have stocked them. A native plant or animal is one that originally evolved in a particular bioregion. The introduction of "wild trout" forever changes any natural watershed. As much as I appreciate the beauty of these transplanted creatures, I wish we could reverse course and give those lovely natives a chance.

Driving in from West Yellowstone we followed the Madison River, which was given this name by the leaders of the Corps of Discovery: Meriwether Lewis and William Clark. Decades later the beleaguered Nez Perce would come through this same magnificent valley while being pursued by the descendants of the U.S. Army soldiers they had once warmly greeted and to whom they willingly rendered aid. What might have been if we had made different choices? Mutual kindness and respect might have led to mutual coexistence and benefit. Imagine a world where we live together. Where rivers run free and clean. Where the air is pure. Where trout rise and birds sing and children play without fear. Imagine.

The Madison River valley is breathtakingly beautiful. Free-roaming herds of American bison wander among the trees and rest in the wildflower-covered meadows. Mountains rise in the middle distance, and geysers send their eerie plumes of hydrothermal steam into the bright blue skies. And then there are the rivers and creeks that sustain wild trout and other wildlife in a manner that seems nothing short of magical or perhaps even divine.

Before fishing, Sue, Josh, and I visited Firehole Falls and the geothermal area surrounding Old Faithful. As we waited for the famous geyser's next eruption, I took a short hike through the surrounding forest while keeping a wary eye out for bears and bison. There were other people walking the trail and so I wasn't truly hiking alone. And there was plenty of sign of both bears and bison along the trail. As potentially dangerous as a sudden encounter might be to my continued existence, I was so pleased to know they were still living here—and thriving.

It was a good thing that I needed to be fully present and deeply aware of my surroundings. I won't pretend that I am fearless when walking in a landscape populated with large apex predators and potentially lethal large herbivores. But sometimes a little fear reminds us that we are alive and that we want very much to stay that way as long as we can. In Nature, it only takes a thousand-pound bear or a microscopic "bug" to keep us humble.

When we reached the Firehole River and went to park the car, Sue waited a moment for another car to pull out before swinging her vehicle into the next parking space in line. As she did, a man in a pickup truck pulled up next to her, rolled down his window, and with great irritation hollered, "You're holding up traffic—move it!" He swung his truck past her and parked hurriedly. We parked next to him. The younger and more foolish version of myself might have allowed his irritation to become my irritation, but these days I am constantly practicing my life skills as an "Imperfect Texan Buddha and Warrior-Poet." I choose how I will respond and do not merely react. There is a big difference. One gives your power away, and the other is empowering. One causes you to become a reflection of your current conditions; the other causes your current conditions to reflect your chosen perspective.

When we stepped out of Sue's vehicle and started stringing up our rods and pulling on our waders, I saw that the grumpy man was already in his waders and gathering his kit to head out for a day of solitary fishing. I looked over, made eye contact, and said, "Good morning." He greeted me in return, and I started up a conversation with him about how beautiful and peaceful the river was, and how fortunate we all were to be here. In time, he relaxed. He transformed from an angry man in a hurry to a gentle man conversing pleasantly with a fellow angler. And instead of launching off toward the river as originally intended, he lingered and chatted with me, Sue, and Josh. It was as if we had never met before under lesser circumstances—and that is exactly what I hoped for when I made the choice to respond and not react.

Just before getting onto the trail that led to the river, I saw a sign that contained various images of grizzly bears, including illustrations of a bear charging a human with the animated figure of a human either "backing away slowly," "playing dead," or "discharging bear spray," depending on the behavior and actions of the bear. The sign also contained the following warning: "Caution: You are in Bear Country. Visitors have been injured and killed by bears. There is no guarantee of your safety." We read the sign and walked onward, heads up, bear spray on our wader belts, and happy to be alive in bear country.

It is worth noting that the man who had transformed from grumpy to gracious walked the trail more or less as we did. At one point along the trail, we came to a small bridge that allowed the river to slip beneath it and us to walk above it. At that crossing of the trail and the river, we said our farewells to our new friend as he walked downriver and we walked in the opposite direction so as to give him the space and solitude we assumed he was seeking. But just before parting, he invited us to fish beside him on the river, if ever we wished. So much had changed in such a short time. The river was calm and peaceful, and so were we.

All around us was wilderness and the steaming plumes of the geysers and springs of geothermally heated water that ran out of the earth and down into the river. In past years we would have been too late in the season to fish the Firehole, as the water temperatures would be too warm and the fish would have moved away from the natural geothermal warmth. But now the unnatural atmospheric conditions have caused the climate to turn upside down, and the unseasonable high, cold, fast water worked to our benefit. The influx of fresh cold water held the trout in the area longer than what might have been "normal" in the pre–climate change past.

The current weather conditions seemed ideal for fishing. We could see trout rising all around us and the beginning of a hatch of midsize whitish mayflies and a coinciding proliferation of white moths that for some reason were also flitting and fluttering just above the surface of the water. I wondered if the miller moths were "hatching" from the muddy banks along the river. At last we had been given useful information during our routine local fly shop visit. We were prepared with good imitations of both bugs, and I tied a fluffy white fly onto my tippet with more urgency than a "Good Texan Buddha" should exhibit. After days of minimal catching and slow fishing with nymphs and bobbers, I was ready for some dry-fly action. If this was sinning, then I was a sinner alive and well in heaven.

It's amusing how quickly I forgot about the potentially dangerous bears and bison once I began to focus on casting a dry fly to rising trout. Just upstream and on the far side of the river there was a small sign that warned us to "Keep Out" due to "geothermal instability," but we had been advised to explore the upstream side, so we decided that as long as we didn't pass the sign we were in fact heeding its warning.

Josh began fishing on the side of the river where the sign was posted, but downstream of the warning, and Sue and I fished on the other side where there was no warning sign. It seemed a reasonable compromise. And it was working, too, because before long I began casting and drifting my dry fly into the feeding lane of a steadily rising trout that took a swipe at it but missed, and Josh was across the river from me with his rod bent toward a nice fish. Sue was quite a ways upstream of me, so I wasn't sure how she was doing. I made another pass at the rising trout, and this time he briefly connected with me, but I was too slow and failed to connect with him. Did I mention that Josh is a better angler than me? He managed to land three fish in the time I managed to miss one.

That was about the time I looked downstream and saw two park rangers standing on the bridge looking at us and chatting. We made eye contact, and the older of the two blew her whistle and motioned to all three of us to come toward her. We did, and as I began to walk from my side the older ranger walked toward Josh, and the younger toward me. I didn't know it at the time, but I was about to meet another new friend. Her name was Ranger Sara, and for once I'm so glad I had become an accidental scofflaw. We met at the center of the bridge and she smiled. I smiled too. A friendship was born.

Sara informed me that they were taking photos of anglers in areas where the signage might be inadequate or confusing so they could make the case for better signage; this place fit that description. I told her that Josh had no intention of passing the sign and that we figured the other side of the river was safe because it had no warning. She said, "No, you're fine. It's just that the signage is poor, and the reality is that the entire area is unstable. I mean, we wouldn't want you to get swallowed up by the landscape, would we?" We both smiled and agreed that being eaten by the earth would be as unpleasant as being eaten by a bear. And that's when Sara Tobin and I became good friends. We shared contact information and I was given an inside view into the life of a National Park Service ranger. It was a life I had once intended for myself—if only life had unfolded as I once intended.

I'm not sure if everything happens for a reason or if we just manage to find a reason for everything that happens, but it turned out to be a good thing that Sara blew her whistle and beckoned us to change plans and fish downstream. As soon as we did, fish began to rise everywhere we looked. Unlike all the other days I had been with Sue and Josh, the problem now wasn't a lack of rising trout but rather an overabundance of fish to target. It was a beautiful problem to work through.

Sue and Josh took the far side of the river, and I fished the near side. Beside me was a geothermal vent that was spewing a steady stream of mineral-rich, superheated water that trickled down the encrusted embankment and into the river. A mix of coniferous forests and meadows surrounded us, with the area nearest the river being meadow. This allowed me to fish while having enough of an open view to become aware of any approaching wildlife in time to make a considered decision about any necessary change in my own actions. In bear country, this was a good place to fish.

The currents were tricky in a few places, so we stood on our respective banks for a moment, watching the trout rise and considering the best approach to the problem at hand. Josh and I hollered to each other about the necessity of longer casts and a bit more slack in our lines than we might otherwise prefer when it came time to set the hook. And that's when he said, "This looks like a place for a 'God save the Queen!' hookset." He saw the quizzical look on my face and immediately demonstrated by quickly raising his rod hand up as if the rod were a sword, then just as quickly thrusting his line hand down as if marching in a comical Benny Hill style, while simultaneously shouting, "God save the Queen!" in his best version of an English accent. I laughed and decided then and there that I had to land a fish with full pomp and circumstance, just for the sheer joy of it.

Fish were rising all along the Firehole, and the pale little mayflies and fluffy white moths continued to fill the air all around me. A few lightly colored caddis flies were coming off as well. Life and death unfolded everywhere, and the only place I did not see rising trout was directly where the heated water was entering the river.

I looked on the horizon where the meadow turned to forest and imagined a herd of bison grazing. I looked over my shoulder into the nearby trees and imagined a massive bear standing there, watching me. Both images were from my imagination. The meadow contained only grasses and wildflowers. The forest behind me consisted of spruce trees that at the time showed no sign of lurking bruins. I felt a mixture of relief and disappointment.

Sue was having a bit of a tough run of luck, but Josh picked up another fish or two and I had a few fish smack at the fly without connecting, then lost another because I failed to take that opportunity to declare myself a loyal subject to my one and only queen—Mother Nature. But there was a cheeky fish that kept rising in a crease in the current, and I decided that she was going to be the one true object of my affections until I managed to land her.

After a few nice drifts without result, I finally enticed a splashy rise and take, and that was when it all came together. While thrusting my hands vigorously apart and loudly proclaiming, "God save the Queen!" I found myself hooked into a beautiful Firehole trout! Josh was across the river watching, and we were both laughing at my antics as the fish jumped and twirled in the air and finally landed on the soft grass next to the river. Yes, I was working without a net, like an acrobatic clown. Wetting my hands I cradled her briefly just below the water's cooling surface, and I thanked her as I watched her swim home. I smiled. It was a good day indeed. And I felt supremely happy.

Fish were caught on the Firehole River that day, but that was hardly the point of it all. I had shared so many wonderful moments with Sue and Josh, and each felt poignant—like the end of something and the beginning of something new. At the end of the day, it felt vital and urgent that Sue and I got the chance to share every last thought that we'd been too busy all week to share. Josh was so exhausted from his constant efforts all week to get us down rivers and into fish that he rightfully took the opportunity to get a little sleep in the back of the SUV as Sue and I enjoyed the scenery and the company on the trip home.

While driving out of the park through the north entrance, we saw a few more bison and also a grizzly sow with two cubs. Wherever there was wildlife, there were traffic jams, and for me this reduced the experience to a sideshow circus atmosphere. Harassed by horn-honking tourists is no way for such noble creatures to survive and thrive. I must admit not having an ounce of sympathy for the idiots who get tossed and trampled each year while disregarding both the regulations and warnings of the National Park Service and the natural space of these wild, free, living beings. All too often, respect seems to be as lacking as empathy. I wonder, why do they even come here?

Sue remarked of her surprise at the unseasonable conditions in the park and the lack of the usual herds of bison and elk along the way. Even the made-for-moose bogs were empty grassy wetlands. No moose. No elk. Damn few bison. We passed areas that were still recovering from intense wildfire, and while wildfire is a normal part of this ecosystem, the frequency and intensity of modern-day wildfires are not. Everything is becoming more extreme. Bob Dylan was all too wise when he sang, "The times, they are a-changin.'" That too felt significant.

We drove through the towns of Emigrant and Livingston and stopped at a few of the spots where scenes from the movie *Mending the Line* were filmed. I sat on the same bench at Angler's West Fly Shop where the characters Colter and Harrison sat while sipping beer and speaking of life. We drove by the bench near the Yellowstone River where Colter and Lucy sat, doing much the same. And as I looked out at these echoes of artificial life imitating what so many of us mistake for "real life," I thought about how many faces have over the years smiled and sobbed while sitting next to the mighty Yellowstone River.

This is our movie. We are writing the script each moment of every day. We can choose to write our character as tragic or brave. We can determine if we will overcome challenges or be overcome by them. We can create ourselves as growing, adapting, determined, bold, humble, adventurous, and selflessly kind, or we can fill our inner dialogue with words of stagnation, ignorance, cowardice, selfishness, and narcissism. I guess the thing to do is wake up every day and ask, "What kind of script do I want to write for my life and my place in this world?"

It was late in the evening when we drove through Paradise Valley toward Bozeman and the house and home of my friends Sue and Josh. And that's when our conversation turned to things that run deeper than what fly to tie on or how we might present that fly or how many fish are caught. We spoke about life.

Sue spoke and I listened as she shared with me the pressure she felt from the expectations of society to keep pursuing promotion to higher and higher levels of management within her workplace. She shared that if she does so, she may have to leave Montana or at the very least be required to travel more often and be gone from Josh, Meatball, Junior, her friends, her church, and her home for extended periods of time. I knew that my friend was symbolically "lowering the tailgate" as we both sought to help each other find a bit of wisdom. So I listened and remained quiet until she had shared it all and in her own time created the pause that told me she was listening too.

I never tell anyone what they should do. Who am I to do such a thing? If I contain any wisdom, it is that within my understanding I understand few things—for certain. But I do try to help people by guiding them toward the most meaningful questions, so that they can discover the answers that they already know. I have found that once we silence the outside voices of society, parents, siblings, peers, and culturally ingrained expectations, questions and answers come to us naturally.

So I asked, "Are you happy here and now?" "I'm as happy as I've ever been in my life," she replied. Then I asked, "Are you doing the things you love to do . . . that feed your soul in the here and now?" "Yes," she answered, "but I'd love to have a bigger positive impact in more areas that I feel passionate about." "So," I continued, "is there any way you can expand your reach and positive influence without changing your job position and thus potentially losing all that you just said you already have?" She confirmed that she could and shared a number of ways she had thought of for doing so. I could tell that one of those paths in particular excited her. She asked, "So what do you think?"

I paused for a moment to choose my words as wisely as possible. I knew this was a moment that called for some vigilant tailgate wisdom. Then I said, "I think it's a fishing lesson." I saw her glance over toward me. "You think it's a fishing lesson?" she repeated. "Yep. You see, we all know this one but sometimes we forget." Then I paused before saying, "You never leave fish to find fish." She smiled. After all, we had the entire Firehole River to fish today, but we invested all our time in one pretty little section where the fish never stopped rising—and we were happy. In angling and in life, we must guard against the compulsion to leave what we love for reasons that have little to do with living an authentic, joyful, meaningful life.

It's important to know when to stay and when to go. It's just as important to understand when change is a positive thing and when you need to change your perspectives, paradigms, and personal choices so as to create a healthier life for you and the world around you. We learn from the past, live in the present, and consider our potential futures; they are not decided in stone until they become our present reality. There is time to respond before we are forced to react. There is time to choose to turn the wheel and go in a new and healthier direction.

As I look at my own physical, mental, and spiritual health, I have come to realize that what I put into my body and what I take out are choices that impact my mortal future. And it's much the same with our relationships with each other and our planet. It's the difference between blindly operating under the illusions of ownership, independence, and "forever after" and wisely choosing actions based on the natural realities of partnership, interdependence, and impermanence. It's a choice. Be a soulless victim of circumstance or a thriving, soulful being that acts with clear-eyed understanding rather than unexamined mythology or unmitigated greed.

I know many of my dreams will never come true. But everything of value in this human world must begin with a dream—that is then acted on. Dreams in themselves are not enough.

We will most likely never choose to change our national parks to a model more like that proposed by the late Ed Abbey where the hotels, concessions, and traffic are moved to the outside edges, and visitor capacity limits are decided by the needs of the natural environment, not the wants of the masses. I am fully aware that I will probably never see the day when bison aren't surrounded by SUVs, or the time when absolutely no one would even consider tossing their beer can into a river. Just as I'm aware it's unlikely that we will ever decide as a people that enough is enough and choose freely to reduce and limit our population, consumption, and impact on the earth—our one and only home. Still, an old warrior-poet can dare to dream.

It really is a fishing lesson. We have been given a beautiful life on a beautiful planet. Why are we so anxious to change it? "You never leave fish to find fish." If we don't choose to turn the wheel, the end result will be the same no matter if it comes to us via the failures of collective human nature or via the collective actions of Mother Nature as she ultimately succeeds in holding us accountable. Beyond my lovely dream of a balanced and peaceful world where humans come Home as part of Nature rather than apart from it, there is the certainty of a far less attractive reality. Sooner or later, every smoker stops smoking.

Chapter Ten

Walking and Wading Teton and Yellowstone Country

To seek enlightenment, intellectual or spiritual; to do good; to love and be loved; to create and to teach: these are the highest purposes of humankind. If there is meaning in life, it lies here.

—George Monbiot

I try to always live in the present moment, however that moment might unfold. I stay vigilant in the knowledge that yesterday's gone and tomorrow may never come, and "now" becomes "then" in a blink. And if you think about it, nothing is real—really. Everything becomes the past all too fast. And while I understand that many people seem to derive pleasure from memory, it more often than not causes me to feel sorrow for the people and places I may never see again. When I see aspen leaves floating downriver in the evening sunlight, I see life drifting around the bend.

Looking and longing backward or forward is a waste of time—whatever "time" might be. But I do tend to experience many flashback memories where my mind sees and feels moments of my past life or lives. I see faces of loved ones and images of rivers, mountains, oceans, and sunrises. They appear in my mind's eye with crystalline clarity and then fade again like drifting dreams. I see the smiles and hear the laughter of souls who have moved on but somehow still live inside me. It's as if these places and people left something of themselves within me. Sometimes it feels as if I left a piece of my soul behind, or perhaps they took something from me that I will never get back. Some of this loss is tragic and much of it is magic, but all of it contains an Irish melancholy that seems to be a part of my literary DNA. Even with the surname of Ramirez, I am one-half Norman Irish. Perhaps this is where I get my proclivity to conquer challenges and then feel too deeply about them. Somewhere in Normandy or Spain there may be a thoughtful angler who shares my same bloodline. I wonder if we share the same smile, laugh, or light in our eyes. There have been millions of smiling faces before mine, and there may be millions yet to come. I suspect that far back in my ancient lineage is a reluctant warrior painting poetic pictographs on the cave wall while wishing for peace.

When my plane floated into Jackson Hole, I was immediately transported to another time when I watched the wing tips of an airliner seemingly skidding just barely past the stony walls of mountains. The first time I experienced this was a decade ago as my Peruvian Airlines jet came in for a landing in the ancient city of Cusco. I will never forget the surreal image of the Andes as they stood rock solid just outside the window of the plane. As we descended I felt myself drifting toward an adventure that almost cost me my mortal life, but one I would never have wanted to forgo.

When I flew into Cusco, I began a journey that caused me to wonder if I would even make it home or if I might die in that forever foreign land. Still, somehow I knew that this was not my place to live or die, just to visit. And now with the stony face of the Teton Range staring back into mine, I felt as if I was somehow coming home once again. Here I felt the desire to never leave this island of seeming sanity in an all too often insane human world. Nature surrounded me as some of the best of human nature awaited my arrival.

I was greeted at the airport by my dear friends Mary and Randall Kaufmann, two of the sweetest souls on earth. Mary gave me a hug as Randall smiled and wrapped an arm over my shoulder. I will never forget that moment and how the fresh mountain air and clear blue skies accentuated the overwhelming sense of joy and gratitude I was feeling. And another flashback came—in a flicker of memory and mourning.

In that instant of breathing the chilled, pine-scented Rocky Mountain atmosphere, I recalled being a much younger man seeing these mountains for the first time as a U.S. Marine traveling with my military brother Bruce. We were driving to Camp Pendleton from the East Coast and decided to camp in the Colorado Rockies along the way. When we first stepped out of Bruce's truck after reaching the mountains and climbing into their welcoming embrace, I was instantly struck by the fresh aroma of pine trees that somehow just seemed "western." Being a Texan I've always been accustomed to living on the cusp of southern comfort and western freedom. Now at last I was feeling what eagles on the wing must feel each day—high above it all, free and full of possibilities.

Randall had been a bit opaque about where we were going fishing the next day. As it turned out, it was one of his secret spots, and I must abide by my word and oath that I will never disclose the location of the Paradise Pool. Like the many places and faces of my Marine Corps days that were deemed "Classified," I will take the location of this most dreamlike water with me to my grave. However, I will share this: It's somewhere in northwestern Wyoming where the buffalo still roam and the deer and the antelope play. It is surrounded by snow-flecked mountains and golden-grassed prairies, and it fills from and empties into a legendary river on a private ranch.

In this case the private ownership protects it, as long as the same person owns the ranch, but like my memories, this too has its limitations. Transitions are inevitable, and the outcomes of each change can go either way in an instant—from restoration and preservation to exploitation and destruction. Grand Teton National Park began as private land owned by good stewards who understood that the only way to protect it beyond their finite lifetime was to make it public land with the stipulation that it be protected for future generations and for the preservation of its native plants and animals. John D. Rockefeller and family led that effort, and had it not been for the determination and foresight of these people and others, this water and wild landscape would almost certainly have been lost forever.

When we turned to go down the ranch road, there was a small herd of pronghorn on either side, with a single buck running across our track kicking up dust. Randall shared that much of the area's migrating pronghorn herd had perished after a particularly severe winter storm left them stranded and starving. With quickly changing global climate patterns, everything is becoming more intense—from drought to flood and hurricane to blizzard.

Paradise Pool is everything it sounds like. I gave it that name as soon as we met. As always, I wondered how many eyes have marveled through the decades and centuries at this sublime landscape where the Teton, Gros Ventre, and Wind River Ranges converge. I held visions of indigenous people camped here, and of French fur trappers and Union soldiers from Iowa, and even the "owner" of this place today. It struck me how we buy and sell pieces of the planet in the same way we sell cattle, horses, sheep, and at one time, people. I wonder if we'll ever learn that we own nothing.

Randall and I began to put on our waders and string up our rods as we watched the river flow by with the evening sunlight shimmering across the water. I was wielding my "Jedi light saber," the 5-weight Orvis H-3 that my buddy Ross Purnell had gifted to me. I love it as if it were an extension of my body, but I don't own it any more than anyone can. We are simply traveling together in this place and time—just like the river. In time, I will pass it on as it was passed to me. Kindness and generosity can be part of our human DNA. Like almost everything else, it's a choice.

Walking through sagebrush is always a viscerally pleasurable experience for me. We do not have sagebrush where I live in Texas, and I regret this, because the sight and smell of it is so western, and so enlivening. If I ever leave Texas, I might follow the scent of sagebrush or the way of tumbleweeds. I've always been a son of the South with a westward leaning.

The riverbed here is covered with slick, rounded stones that range in size from an apple to a cantaloupe, but there's nothing sweet about trying to walk and wade on them, so both of us relied heavily on our wading staffs. Randall mentioned that as we get older, a wading staff becomes more important, but I have the feeling that I could have found myself half-submerged in this river at twenty-six as easily as sixty-two. So I waded carefully into position, just within casting range of a long current that ran through this even longer river bend pool. I couldn't see any bugs on the water or in the air, but every now and then a fish would rise. Each cast and drift felt hopeful.

We were using dry-fly rigs with barbless hooks that we fished as dries but then allowed to swing with the current at the end of the drift. Randall said there was a possibility of a Blue-Winged Olive hatch, so that's exactly what we tied on. Almost immediately I received a surface strike, set the hook, and watched as a beautiful brown trout leapt into the air, shook its head, and tossed the fly back with extreme prejudice. It seemed he was as disgusted by my lousy hookset as I was. Randall didn't make the same mistake, and almost as quickly as I lost that fish he landed one and then another. I missed another that swiped at the free-floating dry fly but finally hooked and landed a nice brownie that took the fly as I let it swing and submerge. As is often the case, there was a brief moment of furious action and then sudden silence.

I don't mind the silent times. They serve so many purposes, effortlessly. When the fish stop biting, it reminds me to be patient, resilient, and curious, and to embrace the moment just as it comes to me, like so many mayflies on the water. And it is during those quiet times that I can pick up my head and breathe in the aroma of sagebrush and wildflowers. It is during these moments that memories are made, in communal conversation and silent solitude.

When the fish refuse to rise, I can watch the waning sunlight just over the horizon and notice the moon waking as the sun slumbers. The crickets begin to sing and the breeze picks up, causing the prairie grass to bend and bow so that the whole earth seems to breathe as I breathe. Most of all, it is during the silent moments that I feel most grateful.

We were working our way downcurrent toward the outflow of Paradise Pool. Randall caught a few more brown trout while I landed a brightly colored rainbow, which he said weren't all that common here. But after a while, when things got most quiet, we traded our mayflies for caddis flies, which seemed to be even more to the liking of the fish of Paradise Pool. We moved back upstream and started working the pool from top to bottom once again. It was big enough to do that—we rested the top while we worked the bottom and then, like life, we went full circle toward our origins. I liked that. It felt heavenly.

My "light saber" was serving me well, and each cast seemed to be landing wherever I willed it to (if you've read this far, you know that this is not always the case). Some days my mojo's working and some days my casting's got the blues. But this evening at Paradise Pool everything felt right. Still, I managed to snag more than one clump of sagebrush and thistle before the night was over. I mean, Jedi or not, it's still me.

Every now and then I would stop fishing so I could cast my mind up into the sky and across the water and over the landscape. In the distance I heard the bawl of cattle and then the bugle of an elk. Above me flew a bald eagle, and somewhere around the bend in the river I heard an osprey calling. I looked behind me, hoping that the pronghorn had returned. Then I looked at the mountains in front of me and began to envision the ambling, ominous meanderings of grizzly bears and the looping, relentless run of wolves. These imaginary visons felt as if they should become real again.

I have yet to hear a wolf howling in the evening half-light. I have pursued that dream for a lifetime, and while it comforts me to know that they once again exist here, it troubles me to hear the hatred expressed by some over their existence. The wolves of Yellowstone country are under siege and, since their delisting, have been slaughtered under the guise of protecting cattle, sheep, and elk so that humans may have the sole right to slaughter and eat them. I understand that life and death are part of Nature's cycle. As a young man I sometimes worked on my friend's family cattle ranch. I've hunted from Montana to Namibia. But I believe we must find better ways to coexist with the living beings that are not currently human.

Feelings are subjective no matter who holds them, but the random slaughter and extermination of apex predators across America feels sinful to me. It feels shameful. It feels wrong. And it is ecologically disastrous to so many habitats, including the waters that trout depend on. I believe we can find a middle ground that allows a place and space for all living things. We can meet this challenge of our own making.

We came to that time in the evening when the light slants sideways so that every blade of grass seems to exist in multiple dimensions and the river shimmers in broken flashes of gold. I envisioned bison walking across the far side to Paradise Pool, but that vision never came true. I imagined vast herds of pronghorn and elk all around me, but they weren't actually there. I felt the presence of something primal and predatory watching me through the willows, but that was perhaps an impression from a past life, embedded in my genetic code. Still, all this was possible here. I guess it's why I felt so at home. I know it's why I felt so alive.

Randall caught several fish for every one fish I caught, and that is as it should be since everyone I fish with is a better angler than me. I'm always learning. But I kept working a riser on a bend in the pool, and after several tries I had her leaping into the air and then swimming from my wet hand back to where we first encountered each other. It was another nice brown trout.

At the top of the pool I noticed a fish steadily feeding along the edge of the current and began to cast and drift over her feeding lane. After about the third try I watched her rise to my caddis, and in an instant we were connected as I set the hook, beginning a battle that lasted for a few minutes of leaping and lunging before I brought a radiant cutthroat to hand, admired her in the water for a moment, and set her free. In the course of a few hours, Paradise Pool had given me a cutthroat, a rainbow, and a couple of brown trout, but most of all, it had given me a sense of peace that goes beyond the description of mere words. Still, I will try to describe it to you.

On Paradise Pool, there was no human death, suffering, fear, anger, hatred, pain, greed, ugliness, or sorrow. Here, I felt only life, comfort, love, pleasure, gratitude, and joy. Here, I felt hopeful that places like this still exist, and that we could together choose to re-create the world we were freely given in the first place. We were given this beautiful green and blue planet to live and thrive on, and all we have managed to do is toss the gift aside and play with the box. But here, I saw something magnificent. I saw the potential for redemption and healing. I saw hope as an action word.

In time, the light grew so dim that I snipped off my fly and simply stood there with the river flowing around me and the evening fading into night. The first stars began to shine, and the moon was growing crisp and clear just above the mountainous horizon. And that's when Randall said, "Well, you've caught brown, rainbow, and cutthroat trout in one evening . . . let's go get you a brookie!"

I followed him through the sage and thistle and wiry wet grass to a kidney-shaped pond between the river and the ranch house. "These fish were stocked here by the rancher," he said. "But you can catch and release one just to have the experience of catching all four species in one evening." I'm not much for catching people's stocked pond fish, but for the sake of the circle I acquiesced as Randall handed me his lovely, delicate fly rod with some sort of topwater thingamajig tied to the end of the tippet. I think it was one of my friend Jack Dennis's purple Amy's Ants, which is a wonderful pattern for so many places. Deprived of my Jedi light saber, I began to cast like a Wookiee and snagged a few sage bushes before finally hooking and almost landing a fat brook trout that flipped off the hook just as the rim of the net touched its autumn-colored body. It counted because I never count the fish I catch or the many times when I catch no fish at all but still manage to have wonderful fishing trips. I'm weird that way. I'm content and grateful, and never was much for using mathematics to quantify a life experience. I prefer poetry over geometry.

I was so happy to simply be on Paradise Pool casting and sometimes catching in synchronicity or solitude with my dear friend Randall. Nothing else really mattered. Everything else was extra. I was going nowhere, fast. Aren't we all?

As we drove down the dirt- and grass-covered ranch road and toward the two-lane asphalt highway toward Jackson, the same lone pronghorn buck ran across our trail once again. I will miss him. And as we drove through the starlit night, my mind wandered in exactly the way I said I wish to avoid at the beginning of this story.

I began recalling the way the light shimmered on the water and across the sides of iridescent fish, and between the waving grasses and blooming lupine. I relived the coldness of the water and the warmth of our smiles, as my friend and I shared every passing moment, which, after passing, always manages to follow us wherever we go. And yes, I let myself long to go there again and live this evening just as it was or as I imagined it, in loops of life that unfold like casts of perfection and purpose. I allowed my mind to wander past the starlight and the moonlight and the headlights of the few other vehicles on the road. My mind drifted toward the mountains that called to me now, like the songs of wolves that I have yet to hear.

I guess the reason I yearn for those wild songs is the same one that makes me feel so uneasy in total silence. Like my lupine brethren, I too call out through the darkness of uncertainty and the unknown. I call out in words written on pages and then wait motionless and breathless for the reply. "I'm here . . . I'm here. Where are you?"

We all ask these lonely questions because life as we know it is a lonesome journey illuminated occasionally by our brief and fleeting connections with others of our tribe. We become more aware and awake as we realize that there isn't any such thing as then, now, or someday. So we keep casting homeward, wherever Home might be.

I guess I'm a quite imperfect Texan Buddha. I struggle and strive for something I cannot define. And in the end there is no end, and each cast leads to the next, and each fish leads to another release, and each release leads to the feeling that I want to be swimming with that fish to wherever he may be going—but this too is silly. After all, just like me and you, he's got nowhere to go and nothing to prove. Life unfolds until there are no more pages to turn and then all we have is our memory, until that unfolds as well and we live on as memories.

I'm at the age when mortality smiles at me each day. It's that same time in life I recall in my father, how he would call out the names of movie stars and musicians that were connected to his youth, just as he realized they were gone. I didn't understand back then in my young, immortal years. But now I see the empty chairs and silent spaces all too well. I feel the urgency to live every heartbeat and breathe with as much gratitude and fortitude as I can manage. I feel the foolishness of human pettiness and petulance. I create hope whenever and wherever I can. I keep casting as I move through the uneasiness of my sixties and on toward the peaceful acceptance that awaits my more mature self. The older I get the more I understand that my life isn't about holding on to anything; it's about healing everything.

Paradise Pool will travel with me always, until my last page turns. And while I will sometimes dream of seeing her again, for the most part I want to remember her as she was with golden grass hair and eyes that shined like leaping rainbows. Still, if I could see her again, I'd like to find her looking like one of those pretty young girls who, through the passing of time and forces of entropy, becomes a beautiful mature woman. I'd like to see her surrounded by pronghorn, elk, and bison. I'd like to feel her clean, cold waters filled with native fish and her skies filled with wild songbirds. I'd like to hear her winds carrying the calls of wild wolves and winged geese in packs and flocks. And I would like to leave her once again while knowing she is safe and sound and not dependent on who or what might "own" her, because she is as she always was.

We own nothing and yet owe everything. With our freedom comes our responsibility. Nothing ever gets done by doing nothing. We can refuse to hurt and choose to heal. After all, we're all just traveling together—briefly. In the flash of living, we can leave a legacy of empathy and respect. We can choose to be kind. And that my friends, is the true meaning of "Paradise."

Chapter Eleven

Floating the Snake River, Grand Teton and Yellowstone Country

The butterfly counts not months but moments and has time enough.
—Rabindranath Tagore

As I write this while sitting at my kitchen table looking out the window into the predawn darkness, I am struck with the challenging and magical life I've led. I find myself pondering the many flashes of memory I hold, which will unfortunately vanish when I do. I guess that, in part, is why I write. It's my way of taking you with me through the verdant cloud forests of Peru, across the crystalline Namid Desert, under the sparkling Caribbean Sea, and over the snowy Scottish Highlands. It is another opportunity to remind myself and everyone who chooses to read my humble words that We Travel Together. How we choose to treat each other and this planet stems from how we treat ourselves.

So many magnificent moments are embedded in my memories, like snapshot photos kept at the ready just under a shoebox lid. I remember the way the sun rose over the Sun Gate of Machu Picchu, and the way the lions coughed and roared at night across the Serengeti Plains. I recall the sweet taste of the spring water in the Abruzzo Mountains of Italy, and the time I discovered the tracks of a brown bear crossing my trail in the snow. And then there was the time I reached the pinnacle of the highest sand dune in the Namid Desert and saw the procession of other dunes as they rolled toward the indigo-tinted ocean. And long ago, before it was surrounded by fencing and tourists, I used to sit in the Roman Colosseum at dawn and watch the sun rise through the porticos. I often had that ancient place all to myself, except for the stray cats that had replaced lions and the ghosts that had replaced the Christians. And now as we launched our drift boat down the Snake River of Wyoming, I knew that I was about to experience another almost forever memory—one that I'd want to take with me and share with you. This is my best effort at sharing the journey with you. It's a shame that "forever" slips away, no matter how tightly we try to hold on. There's never been an ink that can survive the raindrops of time.

The first time I saw anyone floating the Snake River was as a boy watching *The American Sportsman* show with my dad. It was the episode where Jack Dennis guided Bing Crosby and Phil Harris on a fly-fishing float down the Snake River. Whenever those two got together I remember them breaking out into song in between jokes and the absolute feeling of joy they radiated. And now, through the kindness of my friends Randall and Mary Kaufmann, I was stepping into the drift boat of famed Snake River fishing guide and new friend Boots Allen. I should mention that Randall and Jack are best buddies and both men have been kind to me beyond measure. Life had come full circle, from being a six-year-old boy wandering the southern woodlands while dreaming of adventure, to being a man in his sixties drifting into his childhood dreams in the Northern Rockies. I wonder if my dad was with me that day . . . somehow, I feel that he was.

By now you should realize that I'm in love with the entire experience of fishing, and that while catching fish is a bonus, it's not why I fish, really. It's the company I keep and the places I come to know. It's the food, music, culture, and feel of a place, and the discovery that comes with each open-eyed turn, each cast, and every hookset of imagination.

Mary is such a wonderful person, soul, and friend, and a damn good cook. I was so excited to hear that she and I would share this day on the river, and the added bonus was meeting and getting to know Boots Allen. "Boots" is a third-generation Snake River guide, and although he holds a PhD in demography he chose to live his life on the home waters of his heart, as a husband, father, guide, consummate teacher, and, lucky for me, my friend. As a former academic myself, I know that the life choice he made speaks even more to his intelligence and wisdom than any doctoral degree. I wish we'd had more time together. I'd learn a lot more than fly presentation from this man. I'm not sure what I'd have to offer him other than my friendship.

The American Sportsman show I watched as a kid was colorless beyond the personalities of Crosby, Harris, and Dennis. The Snake River we were sharing this day was awash in the many hues of autumn in the Rockies. Aspen turned golden in the higher spaces while cottonwoods shone in sunlit shades of yellow and amber along the river's edge. Lodgepole pines stood straight and proud just beyond the dark green spruce that harbored and hid massive moose, loping packs of gray wolves, and solitary grizzly bears. The Teton Range looked down over the trees and across the water and into the depths of my all too often lonesome soul. But somehow, I never feel alone or sorrowful when I'm in the mountains.

I found myself wondering why mountains call to and comfort human beings, as they somehow always seem to do. For me, however, this is only true when the mountains have been spared the scars of human touch. Add coal or copper mines, ski slopes, condos, or clear-cuts and they lose much of their magic. They become wounded warriors—like me.

Looking up at these jagged mountains, I am struck by my own insignificance. Like a teenager staring at a blemish in a mirror, we humans tend to perceive the world as beginning and ending with the creation of our own faces. But there were billions of faces before mine, and over billions of years the forces of Nature formed these mountains and forgot those faces. Glaciers, earthquakes, and the erosion of wind and rain have shaped this landscape and, ultimately, the living things that reside here. Water moves life and climate moves water, and both climate and water determine how places and faces might look. Adaptation is the hallmark of surviving and thriving.

The Teton Range is one of the youngest in North America but contains some of the oldest rocks on the continent. Almost three billion years ago the metamorphic gneiss that makes up much of the Tetons was formed, and nearly as long ago molten magma oozed up through cracks in the stone and formed the crystallized igneous granite that can be seen slicing through the central Teton Range of Grand Teton, Middle Teton, and Mount Owen. And just above the metamorphic and igneous base of the Tetons are the remnants of an ancient sea in the form of sedimentary sandstone, limestone, and mudstone that contains the fossilized shadows of trilobites, corals, and the shells of other marine creatures.

Global geology puts human ideology in perspective. We are a footnote in the story of our planet. And while glaciers seem to lack free will, we humans seem to lack an appreciation and understanding of the perils and possibilities our choices can offer for the health of our planet, or our own bodies. We can drift or we can row, but we should realize that drifting almost always leads to rough water and a rocky ending.

While wading a river has so many intimate charms of its own, there is something so sacred in rowing and drifting around bends and over slack-water pools and fast-water riffles. It's just so damn poetic and the perfect metaphor for how we must take life. Just like in life, we have no idea what's coming next, and we will never find out if we stand dumbstruck in wavering uncertainty on the shore. If you want to live a life of meaning and adventure, you've got to climb into the boat, push off, and get out into the current—just to see where it takes you. And sometimes you sit silently and allow yourself to float as you make small adjustments with oar or rudder, while other times you choose to drop anchor because you know somewhere deep inside it's time to simply remain still with an open mind, heart, and soul. So you sit and quietly take in all that surrounds you and all it can teach you. In the silence, drifting, and occasional adjustment of choice, you come to understand who you are and why you are here in this particular now.

Just before launching into the river, we noticed a pair of bald eagles resting together within the upper branches of a lodgepole pine. When we pushed off, they flew off down the river as if to warn the fish that we were coming, but I doubt the fish hung around to hear what the eagles had to say. I think we were safe.

Mary kindly offered me the bow while she cast from the stern of Boot's immaculate drift boat. I began casting my dry fly and dropper combo—one of my preferred methods of trout fishing. Boots chose the flies, and his experience and intimacy with his home waters showed in short order as I connected with the first Snake River cutthroat trout and mountain whitefish of my life. I was thrilled about both, and I was taken by the wonderful willpower of the mountain whitefish and how they varied from silvery-blue to bronzy-yellow. Whitefish are a native fish that doesn't get the respect and love they deserve. I'd like to change that.

There was no shortage of promising water toward which to cast, and one by one Mary and I connected with trout and whitefish with some degree of regularity and rhythm. Boots knew all the best places to cast and would point them out to me if I didn't intuitively find them. But I felt I was reading this river well, almost as if I'd been here before. It wasn't at all what I had imagined. Some things were missing that I had placed there in my imagination, while others were present that I never expected. And although I did expect the scenery to be breathtaking, I didn't understand how much so until with each turn I was left breathless once again. Sometimes I just stopped fishing so I could simply stand in the bow and stare at the mountains.

We were catching and releasing fish along almost every stretch of promising water, from the edges of currents to undercut riverbanks to backeddies that had inflows cutting across gravel bars. I was catching roughly two-to-one of Snake River cutthroat and mountain whitefish, and I was never disappointed with whatever came boatside. Each fish was beautiful in its own way. It's interesting how we humans place subjective values on other living beings and different experiences. I'm as guilty as the next person. I'd rather catch a mountain whitefish in its native waters than a "trophy" brook trout in a Rocky Mountain stream. And I'd rather catch a smallish brookie in West Virginia than a "trophy" rainbow in the same waters. Value is subjective. We can choose our perspective.

I also find it interesting that on a day like we were having, when by my standards we caught and released a lot of fish, no single fish stands out in my mind. Conversely, on a day where I fish hard all day only to catch a single fish, I find that I can recall every detail of that one fish and where I connected with him or her. I remember how the fish looked rising in the tiny mountain pool and how I had to camouflage myself in front of a spruce tree while trying ever so carefully not to snag another while trying to make a pinpoint accurate cast, perfect presentation and drift, and timely hookset, landing, and release. Sometimes less is more—but today, I was so happy to be catching fish.

Rounding a bend and blowing through a set of easy rapids, I noticed the sounds of elk bugling just beyond the tree line and I kept my head up in between each cast, hoping to see them. Boots saw a bull walking through the trees, but by the time I looked he had vanished. We were just about to give up the search when we came across another drift boat and the guide asked, "Did you see them?" We replied, "See who?" "A whole herd of elk just crossed the river in front of us," he replied. We hadn't seen them, but on the next turn we spied a big moose cow with a half-grown calf, both standing half-submerged at the river's edge. Then, around another river bend we watched a delicate-looking mule deer doe browsing leaves where a meadow met the river. We never did see any wolves or grizzly or black bears on this float, but we could have, and in that possibility there is magic.

Mary and I were both catching fish and took turns cheering for each other and admiring the fish briefly before Boots quickly released them directly from the net back into the river. We never touched a single fish that day, and it felt right. We practiced grinning without gripping, and after considering the respect shown to these most beautiful of fish I came to an almost spiritual revelation: I will no longer pose with fish held triumphantly from the water. From this moment on, whenever I feel compelled to capture the moment by way of photography, it will be while releasing a fish from the fish-friendly net or at boatside as directly as possible back into the water, causing the least amount of trauma to the fish.

This doesn't mean I will never choose to kill and eat another fish—quite to the contrary. I am an omnivore in the circle of life, and some fish are either unnaturally overpopulated or unnaturally introduced into a watershed. In those instances, my predatory nature is both potentially beneficial and completely natural. But I've never felt completely at peace with the idea of causing potentially harmful stress to another living being just so I could count coup and have the experience. Still, it feels important that I connect to these creatures and their home waters so that I might become a voice for the voiceless. And whenever I have chosen to kill and eat a deer, pronghorn, or rainbow trout, it has been done with respect for the life I am exchanging for mine. In the twenty-first century, we need to be enlightened anglers, hunters, and hikers. Our footprints have grown too large for anything less.

The Snake River cutthroat trout is certainly one of the most stunningly colored and patterned fish I've ever known. Because I am slightly pink-and-blue-green colorblind, I often wonder how much I am missing when I look at wildflowers, songbirds, and leaping trout. Still, it was easy for me to see the golden-yellow sides of these handsome fish and how they were covered with what seemed like a million finely placed spots that grew to a crescendo of intensity near and on the tail. The darker but shimmering slate-colored back perfectly offset the warmth of the brightly lit sides and belly of each fish.

The Snake River or fine-spotted cutthroat trout is a natural work of art that currently survives in one small corner of this imperiled planet. When the first Euro-American mountain men arrived, these gorgeous living jewels were swimming between the glacial dams that had allowed them to evolve from the Yellowstone cutthroats that live both above and below their rarefied home waters in the shadow of "Les Grands Tetons." I wonder, did Jedediah Smith, Jim Bridger, and Davey Jackson ever notice how beautiful their dinner was, just before it entered the campfire frying pan?

Today, Snake River cutthroats have been transplanted outside their native range as the result of bucket biology and the pressures of the free-enterprise angling economy. But I, for one, want to find these magnificent creatures right where they belong, between Jackson Lake and the Palisades Reservoir in the Snake River and its major tributaries. I suspect that the future of humanity and this planet will reside not in what we change and re-create as much as by what we heal and rejuvenate.

We are all guests on this beautiful, wounded world that we call home. Why is it that we more often than not act like troublesome renters rather than a community of residents? That's no way to treat someone or someplace you love. There's no place like home. It seems to me that as long as we stand around waiting for heaven, we will continue to make this our hell. Why would we choose that highway when the off-ramp is within our grasp?

I could hear more elk calling out their songs of want and warning just beyond the river's edge and behind the screen of cottonwoods and aspen. The night prior Randall and I sat watching a young bull chasing away an even younger bull before returning to his other pastimes of herding cows and thrashing willows with his magnificent antlers. I enjoy watching wildlife and gazing at mountaintops as much as I enjoy catching fish, but in truth, I didn't want to stop catching and releasing these amazing Snake River cutthroats. I didn't want to stop chatting with Boots about his many fascinating journeys or smiling and laughing with Mary as we reveled in the journey we were currently sharing. I didn't want this float to end, but everything does, no matter our desires to the contrary.

We had launched from a place known as Deadman's Bar and were only a few bends in the river from our takeout at Moose Landing. Boots pointed to a line of current up ahead and said that it would be our last chance for another trout before the takeout. I was in no rush to get there. I really didn't care if the biggest, brightest fine-spotted cutthroat trout in the river was waiting there, just for me, but instead kept hoping it would take us some time to float from where we were to where Boots was pointing. It didn't, and as Mary and I both cast into the seam of quick water, almost poetically, one after another, we took turns catching and releasing our last trout of the day and, quite possibly, the last Snake River cutthroat of my lifetime. When Boots flipped him from the net to the river, I felt rejoiceful for the fish and a little remorseful for myself. I'd felt childlike and full of life throughout the entire day. I felt fortunate for every breath in my lungs, beat of my heart, and bend in the river. I knew it was the last bend in the river—for me.

I realize that even now in the late autumn of my life, I am still a child with much to learn. I'm at that point in a man's life where metaphorically my back cast is longer than my forward cast. A man in his early sixties can easily find himself breathless at the prospect of aging's accumulating challenges, which in my case take the form of genetically driven heart disease, evolving asthma, sleep apnea, and hands that aren't as nimble and eyes that aren't as sharp as they once were. It's all too easy at age sixty-two to wonder if you have the life expectancy of a cocker spaniel and worry that you need to get on with whatever you want to see and do before it's "too late." But fly fishing and my most mature fly-fishing friends teach me almost everything I need to learn. Fishing teaches me to be present and patient and to adapt to the conditions I find around every bend—in a river or in life. My friends teach me what they've already learned by turning the corner, and finding more corners. When I share time with my more enlightened tribal members, I see that they're not counting years or considering yearnings. They are living life every moment of every day—come what may. Come to think of it, there's a lot I can learn from that cocker spaniel. I'm not growing old, just older.

Part IV

Classic Waters in the Commonwealth—Pennsylvania Spring Creeks

Chapter Twelve

Wading the Upper Little Juniata

I live my life in widening circles that reach out across the world.
—Rainer Maria Rilke, *Book of Hours*

I remember walking into a bookstore and picking up a copy of *Fly Fisherman* magazine and saying to myself, "Someday, I'd like to write something and have it read from the pages of this magazine." That seems like a long time ago, almost as if it were in another lifetime that I once lived. But lifetimes are measured in moments that matter, and one such moment came when Ross Purnell, the publisher/editor of *Fly Fisherman*, reached out to me to share that he had read my first book, found value in it, and wanted to invite me to write an essay for the magazine's "Seasonable Angler" column. I was all too pleased to do so, and after it was published, he asked for another, and then another, and then arranged for us to have a phone conversation that led to one of the great honors of my life. He asked me to take over the vacant position as full-time writer of the column—the same column that was originated by an author whom I have respected and even revered all of my adult life, Nick Lyons.

Sometimes, I feel this was predestined. Even early in my life I felt compelled to write, and I recall looking at the "back page" essays in outdoor sporting magazines and thinking, "That space is the perfect space for me." Sometimes I feel as if I may not be the one writing these books and essays as much as I am simply a conduit for delivering hopefully timely messages to where they might be so desperately needed. After all,

I don't really write about fishing; I write about living. I write about the need for cultivating mindfulness, empathy, kindness, love, and resilience in everything we do. I write about balancing liberty with responsibility, and about replacing personal suffering with the innate ability to adapt, learn, grow, and thrive through every hardship and challenge. I write about acceptance over tolerance and understanding over anger and fear. I write about the journey for self-awareness, so that we can become increasingly aware of the limitations of seeing ourselves as "separate" from all other life on the earth. Ultimately, I write about Nature and the best of human nature, while not looking away from the darkness within us or the destruction it can cause. And I am doing this with a keen sense of urgency because I know that my broken heart will get only so many beats, and my asthmatic lungs will get only so many breaths. If there is any meaning to our lives, it's the meaning we choose to give it by how we treat each other and every other living being—including the earth itself.

So much time has passed since that first phone call. As I stepped off the plane in Harrisburg, Pennsylvania, and met Ross for the first time, it felt as if we had finally arrived at the place in time and space that the universe had demanded. We were two friends, sharing moments in a brief whisper of a lifetime, trying to make sense of it all and to perhaps do a little "good" along the way.

We had never met in person before this moment but had already built a lasting friendship by turning social media, email, and FaceTime upside down and ensuring that the airwaves and algorithms worked for us and not the other way around. Every tool, from a scalpel to a spatula, has the capacity to heal or harm, depending on the choices of the user. We can use our words and our works to drive us apart or bring us together. It's a choice we make. The outcomes of these choices are predictable.

The drive through the Pennsylvania countryside was peaceful, hopeful, and miraculously verdant. Forests and farmlands tumbled over rolling hillsides and ancient low-slung mountains. Fields of corn, beans, and grain gave way to woodlands of oak, maple, hickory, and hemlock. For Ross this place was home. For me it was the homeland of my father. It was a "home place" that I was seeing again, for the first time. I could imagine my father as a young man, home on leave from the U.S. Air

Force, walking these farm fields with his shotgun in hand and his beagle beside him. I could see him dapping a small stream for brook trout and drinking spring water from his cupped hand—imagining himself as indigenous rather than immigrant. My father and I always felt the same respect for the enduring echoes of this land's first peoples. "Discovery" is all too often another word for destruction.

Not far from the place where we'd be fishing is the Carlisle Barracks, where the children of indigenous tribes from across the nation were taken from their families and forced into assimilation at the government-operated "Indian boarding school." Once there, the male children had their traditional long hair cut off, and both boys and girls were given new European names and dressed in the clothing of the Euro-Americans. They were subject to corporal punishment for using their native language or in any way reverting back to native traditions. Although children were sent here from as far away as the Dakota and Arizona Territories, locally, the children of the Haudenosaunee (Iroquois Confederacy) were compelled to leave their families and attend this new American reeducation camp. Over ten thousand indigenous children endured this experience, which was established not long after a war was fought that led to the emancipation of black African slaves. So much suffering once took place in a town that now has a lovely trout stream where anglers rest on nearby wooden benches and kids feed breadcrumbs to ducklings.

As farms and woodlands slipped into view and out again, I began to ponder all the human history that has unfolded in this currently pastoral landscape. Gone are the old-growth forests, eastern elk, bison, cougars, and gray wolves. The Amish replaced the Iroquois and retired factory workers from Philadelphia begin to replace the Amish. It's important to realize that every time and place is what we make of it. As we drove toward our destination I wondered, "What will we make of these times and this place we call 'America'?"

In time, our drive brought us to the Little Juniata River, flowing freely just over the guardrail of the highway. We followed the river in the direction of its source, through the tree-lined winding country roads and finally to the confluence of Spruce Creek and the Little Juniata—our destination. The small rural town of Spruce Creek felt just right, a home away from home, whatever that might mean.

After stowing our gear in the rooms of our Airbnb, we grabbed a couple of beers from the cooler and sat on the porch to raise a toast to the coming adventure and our friendship. I chose a local brew called Sunshine Pilsner. It seemed appropriate for the occasion. Wherever I travel I love to taste the wines or microbrews that are locally crafted. Food, drink, and music are three important portals into any culture, and I always want to immerse myself in a culture wherever I may wander. I seek to be a traveler, not a tourist. There is a difference.

As the evening light waned and the first starlight began to wink in the darkening sky, we walked down the dirt road to what seemed to be the only eating establishment in town, the Spruce Creek Tavern. As we walked into the tavern a sign on the door read, "No Waders Allowed in the Dining Area." Inside I absorbed the ambience of the room, which consisted of wooden walls adorned with a few ridiculously tiny white-tailed deer antlers, a couple of metal signs with sporting art and the brand names of Winchester and Remington emblazed on them, and an old crosscut saw that hung beside the painted words, "Today I will be happier than a bird with a French fry." The Outdoor Channel was playing soundlessly from a big-screen TV mounted on the wall behind a podium where a young man with long curly hair, blue jeans, and a plaid shirt acted as the maître d' when he asked, quite formally, "Two for dinner?" After confirming that there were indeed just the two of us, we were seated at a high table next to the ladies' room and greeted by the bartender-turned-temporary-server who had the air of a young woman on guard for older men wearing camouflage and acting in predatory mode. She was pleasant, professional, and aloof. And she was kind enough to bring me a second local brew that I simply could not pass up—Spruce Creek Ale.

Looking at the menu I discovered that the tavern advertised itself as "The Home of the Famous Fries." Prior to my discovery of heart disease and a coronary birth defect that acts to restrict the amount of oxygenated blood flowing into my aging heart, I might have welcomed the tavern's menu with its deep-fried this and that and everything slathered in cheese, sauce, and salt. And of course meat, meat, and more meat on every plate, including the salad (no substitutions). As it was, I decided to surrender to its charms and simply absorb its color and cholesterol.

Even though I made my best effort to find something on the menu that was heart healthy, I was forced to succumb to a "standard American diet" that I had previously known and all too often enjoyed. Sometimes we just have to adapt and hope for the best, but once again my travels reinforced the notion that American eating habits are the leading cause of early disability and death. We really need to change the way we treat our inner and outer environments.

How we choose to treat our bodies is no different than how we choose to treat our rivers. We can allow them to be replenished in natural spring water or absorb acid rain and industrial waste. Depending on our choices, we and our rivers either live or die. That said, I went to sleep that night in the dark silence of my room with only the sound of my heartbeat to keep me company. It was a good night because I slept well and eventually woke up to find that my heart was still beating. Every morning we wake up is a beautiful morning.

Dennis Pastucha serves as the artistic director of *Fly Fisherman* magazine. He is also a great guy and now my friend. I was thrilled to hear that Dennis was driving up from his home in south-central Pennsylvania to fish with us that day, and when we met up along the banks of the Little Juniata it was all smiles and handshakes and exclamations like, "We meet at last!" After introductions and stringing up rods and pulling up waders, we wandered through thick and heavy brush and briar and across lush, green forest edges that were covered in poison ivy until we eventually reached the river. I happen to be quite reactive to poison ivy, and as I walked behind Ross and Dennis through patches of it that I would never have attempted to penetrate had I been alone, I marveled at the courage—or foolishness!—I was willing to show when in the company of friends. I also hoped that my waders would protect me from becoming a red, swollen, blistered mess. They did.

This portion of the river was not our true destination for the day, but since we had decided to meet up here we thought it might be prudent to give it a try before moving on. The river was narrow in this spot, with slick, round, treacherous rocks that seemed to be begging for an ankle to break, but I did my best not to oblige them. I also tried my best not to fall in and look like a clumsy idiot in front of my friends, and this time I managed to remain upright. Past experience has taught me that remaining upright is not a given.

Like everywhere I'd been, from Alaska to Montana and Wyoming, the top half of the country was experiencing unseasonable rains and flooding while parts of the lower half were drying up. In my beloved Texas Hill Country, several of our spring-fed rivers had gone dry to the limestone, and the carcasses of fish and other aquatic life were lying desiccated and lifeless everywhere. But here the water was unseasonably high, fast, and cold with no mayfly hatch expected until the evening. Due to the conditions we were obliged to nymph and bobber fish, which, while effective, is not elegant. And it worked, with Ross catching the first fish and all three of us getting into a few before we decided to wrap it up, pack up the vehicles, and move on to our true destination. I'm so glad we did.

I rode with Dennis in his truck, and we chatted about life and his life at *Fly Fisherman* magazine and how well he and Ross worked together. Earlier in the day Ross had referred to Dennis as a "lifesaver," and I felt obliged to share that comment with him. And I made no secret of how much I value and appreciate being a small part of the family and how seriously I take my humble contribution to the effort. Yes, I know it's a business, but that's not what I see driving what we're doing; we just need to be aware that the goal is to provide what readers want and, even more importantly, to my mind, what anglers and outdoor enthusiasts need to know. I love the balance between the "how to" and the "where to" and the "why" of it all. And slowly, almost imperceptibly, I see us asking the most important questions about our role in caring for our outdoor community—both human and nonhuman. I feel good about that.

As is so often the case, the best destinations take the most effort to reach, and this one was no exception. I won't tell you where it is, but I will say that it was at the end of the road and surrounded by protected forest, and entailed hiking a few miles up and down a woodland trail before climbing down an embankment, through more trees, brush, and poison ivy, and eventually stepping into the river, which here was wide, fast, and wild. I was in love.

The wading on this portion of the Little Juniata reminded me of one of those outdoor challenge game shows where the contestants had to walk across a gauntlet of giant floating foam balls while someone on the sidelines shoots at them with a water cannon. With this in mind, I decided to deploy my wading staff and never regretted that decision. At five foot six I had to work a bit harder than my tall and long-legged buddies who seemed to effortlessly step over fallen trees and half-submerged boulders that caused me to have Marine Corps boot camp obstacle course flashbacks. With that said, I was deeply taken by the beauty of this stream and the surrounding hardwood forest with no indication of civilization besides the occasional sound of a passing train on tracks that were hidden just behind the trees, somewhere beyond the opposite riverbank.

Birds seemed to sing from almost every treetop. Hardwood trees reached high into the sunlight. The river felt healthy and alive, and I was so grateful to learn that both deer and black bears still thrive in these forests. There was a time when both were extirpated from Penn's Woods. We can do nothing about the wounds that humanity may have caused to people or places in the past. But we can do much about healing those wounds in the present, and making things whole again.

Each of us spread out and began drifting our nymph rigs along likely places. At first, there were no bugs on the water, the sun was high and warm in the clear skies, and although the weather was gorgeous it was not giving us either rising trout or insects. Ross and Dennis both caught a few fish and eventually I managed to land one, but the pretty day was, so far, pretty slow.

This is a good time in the story to mention the magnificent fly rod I was casting. It was a prototype four-piece bamboo 4/5-weight crafted by my friend Jerry Kustich of Sweetgrass Bamboo Fly Rods. Jerry is, among other things, an angler, author, and gifted bamboo rod artisan. He is also a hell of a nice guy. Before taking this trip I had been talking with Jerry about my interest in using the original and organic substance of bamboo for at least part of this journey, and he immediately offered to allow me to take his new four-piece out for a spin and only asked that I give him feedback on its performance. Well, in short, it was lovely, soft, delicate, and felt like a labor of love in my hand. His generosity in lending it to me made me a little nervous. I felt a bit unworthy, and also had that nagging thought in my mind as I stumbled across the slick-rocked river and wandered between the willows: "What if something happens to this beautiful rod?" As I cast it I wasn't sure if I should allow myself to feel ecstasy or trepidation. I ended up experiencing both.

Over the next few hours, we each took a section of the river and patiently lobbed our nymphs into the currents, hoping for the best. And we all caught a few fish, although the catching was far from fast or furious. At one point Dennis switched to swinging a wet fly, but I'm not sure it made any difference. And as much as I was enjoying the poetry of casting that fine bamboo rod, on several occasions the rod tip came off and I was forced to reel it in and reattach it. It was so delicate that I was fearful of pushing it back on too firmly, but I also was fearful of losing it. Luckily, I had a line running through it with an indicator and nymph rig at the end of the line. Otherwise, I would have certainly lost that precious rod tip to the winding currents of the Little Juniata. It was unnerving.

It seemed that everything was waiting for the earth to spin far enough to sink the sun below the western horizon. Only then, when the light had faded and the temperature dropped, would the mayflies rise, bringing the slumbering trout with them. And since things had been slow, we decided to get into position with about fifty yards between us in the long, deep pool just upstream of the train trestle that crossed the river.

I could feel the air growing colder and the light growing dimmer, and I decided that I had enough of flinging and sometimes tangling that clunky nymph rig with my borrowed wisp of a bamboo rod. So I decided to tie on a sulphur dry fly and then calmly and patiently wait—along with everything and everyone else. Nature guides us if we listen. Some of the most meaningful moments of my life happened when I waited and watched with an open mind and heart—expecting nothing and accepting whatever unfolded. This is the essence of being a mindful angler.

It was a magnificent place along the river to just stand in the water, feeling the comfortingly relentless currents pushing steadily against my legs . . . and simply wait. It was peaceful, and so was I. Sights, sounds, and sensations are best absorbed as we do nothing. Paying attention is always a solid investment.

And so I stood there in the river, my eyes beginning to strain to see the few mayflies rising and my body beginning to shiver in the deepening cold. The sounds of birds singing their evening songs were punctuated by the occasional whistle of a passing train. I didn't mind the trains at all. There was something magical and memorable in the chugging of the engine and the clacking of the tracks and the lonesome moaning of the whistle blowing in the distance. Images flashed through my mind of men without means catching free rides wherever the tracks might take them. Bluegrass songs came to mind with lyrics that spoke of longing, searching, wandering, and hoping for something between freedom and the stability of finding home. And then I thought about the trains of the Fitchburg Railroad that ran through the woodlands of Walden Pond, and how Henry David Thoreau would sit on the doorstep of his one-room cabin and listen to its whistle "sounding like the scream of a hawk sailing over some farmer's yard." The whistle of the trains that "touched" the "Little J" did not sound at all hawkish to me. It sounded both lonesome and comforting, as if it was reminding me that I did not belong in this river, but that there was a path to and from wherever here and there might be.

It was much the same for Thoreau back in 1853 as he walked the track that "touched" his pond, about one hundred "rods" from the place of his humble cabin. If you've ever been to Walden Pond, as I have, you know that even then, Henry David did not live far from "civilization." And according to his writing of the time, he frequently walked the railroad tracks to travel to and from the town of Concord. While I stood there shivering in water that was growing colder by the moment, I thought of how much I had wanted to walk the tracks to get here, and how much I hoped we might walk them this evening, when it was time to return to the end of the road.

And then I wondered why it is that I regretted walking the woodland trail instead of the tracks of the railroad. Why would it matter to me, the woodland poet who so often shuns concrete and steel? Perhaps it is the romantic in me, the secret hobo traveler who's spent so much of life doing what was expected and not enough time doing whatever the hell pleased me. Or perhaps it's because I have to come to terms with the reality that I am not a city mouse or a country mouse; I'm both—and neither. That's a tough epiphany for an Imperfect Texan Buddha. I'm not the wild man I might imagine myself to be. I enjoy comfort.

I looked downstream to see Ross standing about a hundred yards closer to the railroad trestle and then upstream to see Dennis, who appeared to be edged in a golden glimmering halo made of the day's final rays of sunlight. He looked like the patron saint of mayfly hatches. And as if he had created a miracle by simply standing there in the waning daylight, I noticed what seemed like thousands of bright sulphur mayflies riding the currents of the river and rising into the currents of air as far as my aged eyes could see. It was a miracle, if ever I've seen one.

Just as quickly as the mayflies appeared, the previously unseen trout quickly followed. Just moments before, the river flowed unbroken and silent and seemingly nearly empty. Now the splashing, smacking, slurping sounds of feeding trout seemed to overwhelm the singing of birds or the whistle of any passing train—and in unison, we cast. The soft bend of that Sweetgrass rod sent the little sulphur fly across the river and into the feeding lane with more grace than I thought possible. And it took only a cast or two before a trout took the fly, and I coaxed the trout gently but persistently out of the quick water and close in to my place in the river. He came to my hand and I held him briefly in the water, slipped out the barbless hook, and whispered, "Thank you." It felt good to watch him swim away. When I looked up at my friends again, I could see that both of them had rods bent and fish on. They seemed happy.

I paused, just for a moment, to cheer for my friends, then I took a deep and grateful breath as all around me trout were rising and mayflies seemed to cover the water and fill the air. The transformation was so sudden and so complete that it felt supernatural—but, of course, it's completely natural. It always amazes me how invisible fish and other wildlife can be when they want to be, and how visible they become when either food or sex are the main things on their minds. It occurs to me that human men are quite the same. It's incredible how quickly just about any man will lose his mind over a pretty girl or a pepperoni pizza. Fluttering feathers, flashing fins, and flexing muscles all dance to the same biological music. We are helpless in its primal rhythm.

While I was casting and catching fish after fish to the front of me, I kept hearing thrashing rises behind me, and one was so close that it splashed a bit of the river across the back of my neck. I got the message and turned around, at first watching to see where the feeding lanes were and then casting to them. There was one trout in particular that I wanted to target because she was so aggressive with each steady rise. She went for my sulphur dry fly on two occasions, but I was too slow and missed the hookset. The third time was the charm, and in no time she was in my hand looking beautiful and unamused. I set her free with my usual words of gratitude.

I'm convinced that gratitude is a key ingredient to a happy life. And every moment of every day I seek out the many reasons to be grateful. Yes, I was grateful to be catching and releasing these beautiful fish, and for the epic sulphur hatch we were all experiencing. I was grateful for this soft, glowing evening that was quickly turning into a clear, cold night of star-speckled skies. I was grateful for my body's ability to shiver as the cold sank deeper inside me, and the bracing chill of the night air reminded me of my mortal limitations. And perhaps most of all, I was grateful to be here with my friends Ross and Dennis. I was so happy each time I saw them connected to yet another fish, and to see all of us in that moment—joyfully childlike and forever young. And we all caught fish after fish until circumstances beyond our control told us, enough is enough.

There are many ways that the universe can tell us that we are done with any journey. It can be the shivering from the cold or the thirst from the heat. It can be a hunger in your belly or an ache in your back or a twist in your ankle. In the best of times it is simply that you have the spiritual maturity to say, "I've been given more than enough today, and it's time to let go." But today it was none of these for me, although I was shivering in the cold, and hungry from the long day, and my back was hurting—just enough. What did it for me was a fish, a flub, and a foolishness that I immediately regretted. Let's start with the fish.

The sunlight had vanished, with only the slightest, almost ethereal glow illuminating the river around us. It was just enough light to see the trout rising but barely enough light to see your fly line in the air or on the water. What I could clearly see was a good-size trout that was slashing the surface with big aggressive strikes, one after another as the mayflies drifted into his feeding lane. Even Dennis could see the fish from where he stood, and he commented on it by saying, "That's a nice fish." As we all know, "nice fish" is angler-speak for "big fish." I decided that this fish was going to be my final target—and he was.

I began casting in the dim light into his feeding lane. He ignored me twice, and on the third try my cast was short and led to me landing a much smaller, but still quite pretty little trout. After quickly releasing the youngster, I watched to make sure he was still rising, and he was. So I set up my cast with as much skill as I could muster with that soft, sweet bamboo fly rod that I still wasn't quite used to and watched as the fly landed like an autumn leaf, floated true, and was engulfed in the eager mouth of that magnificent trout. I set the hook, and the fish twisted and turned and bore down into the depths toward some submerged branches, and in a moment of adrenaline-induced idiocy I tried to horse him away in my usual Texan bass angler style—and broke the tippet. Not only was I heartsick at the loss, but I also felt all of my Catholic school guilt hit me at once and could almost hear Sister Mary scolding me about the fly I left hanging in the lip of that "nice fish." That's when everything came undone.

As I gathered up my line and prepared to try to tie on another fly, the right lens of my glasses simply fell out and plunked into the river. On top of that, the cold had caused my arthritic hands to stiffen, and I couldn't stop shivering. I tried closing my right eye and using the glasses like a monocle, but between the dim light, half-blindness, and shaking stiff fingers, I could not tie a new fly on no matter how hard I tried. It seemed the universe was telling me, "You're done." So I reeled in and waded through the rising trout toward my buddy Ross, who also was reeling in as he called out to Dennis that it was a good time to call it a day. Dennis began walking downstream toward us, though even from a darkened distance I could tell that he wanted to keep casting.

I was happy to hear that Ross intended for us to walk the railroad tracks back to where our vehicles were parked at the end of the road. There is something so "Americana" and romantic about it in my mind: three friends traipsing down the tracks under a blanket of starlight after a day of adventure. But to get to the tracks, we had to wade through a muddy bog and then climb over a huge heap of fallen timber and up a hillside of rocky scree—in the dark. But we made it and walked quickly across the trestle while listening intently for oncoming trains. A pungent-smelling deer carcass on the tracks emphasized the importance of vigilance, and the aroma of its death hung in our nostrils for most of the journey—just in case we needed to be reminded. I guess it wasn't all romantic.

At one point we had to bushwhack a short way from the track to the trail and then down to the parking area where we began breaking down our rods and stowing our gear within the beams of our headlamps. That's when, much to my horror, I discovered that the rod tip of that beautiful, handcrafted bamboo rod was missing. And there wasn't anything we could do about it. That last section of the rod was as thin and delicate as a feather, and from the time I had lost the fly and was unable to reattach a new one, I had waded through a few hundred yards of current, slogged through a grass and willow bog, climbed a debris pile and a hillside, walked a couple miles of train tracks, and wandered through the forest—in the dark. I had to accept that the rod tip was gone for good and I owed my buddy Jerry an explanation and a check made out for the total value of the beautiful rod he lent to me. These are the times that test my Imperfect Texan Buddha spirit, but I was up to the test and I'm grateful.

Now, as I sit here at my writing desk in my Texas Hill Country home, that moment seems like a long time ago, almost as if it were in another lifetime that I once lived. But lifetimes are measured in moments that matter, and one such moment came when I looked over at Ross and saw how he was sharing my anguish over the loss of that lovely rod tip. I reminded myself that "the die has been cast" and there was nothing I could do about what had happened. But I could choose my response and how it impacts my friends, our friendship, and our time together. I looked over at my friend and said, "No worries, Ross. I will not let this ruin our trip. We will fish Penns Creek tomorrow as planned, and I will call Jerry and tell him what happened and offer to pay whatever it costs to make this rod whole again."

And I did call Jerry and with a heavy heart recounted the entire series of events and asked him, "How much do I owe you?" Jerry laughed, "You owe me nothing," he said. "You were testing the rod for me and did exactly what I wanted you to do as I perfect the design. Send it back to me; I will reconsider the location and design of the last segment and send it back to you for more testing . . . if you're willing." I was more than willing. I was honored.

I'm no longer a member of any organized religion—although I find value in the language of many of these ways in which humans attempt to describe the indescribable and know the unknowable. In this case I am drawn to a Yiddish word that fits this situation well. My friend Jerry Kustich is a true mensch!

As we stood there under the starlit skies with only one more choice to make, I felt so fortunate to share this day with Ross and Dennis and the Little Juniata. As for the choice, while we were driving back, Ross told me how he had made a special batch of his homemade brisket in honor of my visit—all the way from Texas. He told me about the care and process he used in its preparation, and I could hear the humble pride he took in his cooking. (As you might recall, I have been excluding meat and most animal products from my diet since my diagnosis of heart disease. Ross made the brisket before he knew of this.) When we lowered the tailgate and opened the cooler, we each grabbed an ice-cold brew to celebrate our day together, and that's when Ross laid out the brisket, bread, and pickles and said, "You don't have to try it . . . I didn't know." Through the darkness I smiled as I reached for the smoked meat and loaded up the pickles and said, "Well, as my cardiologist once said, 'Ya gotta live!'" We all laughed, and I have to admit that just thinking about it now makes me hungry and causes me to smile. And one more memory about that night makes me smile. When Ross asked me if I was fine with us moving on that night and fishing Penns Creek in the morning, I said, "Well, yes, on one condition . . . can I borrow a fly rod?"

Chapter Thirteen

Wading Upper Penns Creek

The best day of your life is the one on which you decide your life is your own. No apologies or excuses. No one to lean on, rely on, or blame. The gift is yours—it is an amazing journey—and you alone are responsible for the quality of it. This is the day your life really begins.

—Bob Moawad

In the morning, Ross and I woke to the sounds of birdsong and the occasional screeching call of blue jays. People in the northeast are probably so used to blue jays that they pay them no mind at all, but at home in the Texas Hill Country, I have only ever seen two blue jays and they were both in my backyard, ever so briefly. In fact, it may have been the same bird, simply passing through while migrating somewhere to the north, on two separate occasions. When I saw him, I gasped in surprise and delight. I think people who see them every day forget how blue they are and how bright they are in both color and intelligence.

We all fall into this trap. I was reminded of this while walking in my favorite little Texas canyon park with my wife, Alice, our daughter, Megan, and her partner, Nick. When a bright red male cardinal flew up into the Texas mountain laurel, Nick reacted much as I did to the blue jay. To him, the sighting of that bright red bird, which is unknown in his native country of England, was a special moment. Each day, I do my best to see the world through childlike eyes. I never want to lose my sense of wonder and my desire to wander. I never want to lose my penchant for getting lost so that I might find something surprising and miraculous.

We need to guard against the habit of treating extraordinary things as ordinary. This includes each morning when we wake. Waking is never guaranteed. Gratitude is appropriate with each new day and every time a flash of color lands in our lives, even briefly. Sometimes that flash is on the wings of a bird or in the laughter and light of a friend. Nothing in life is ordinary. Everything is a miracle. Everything is full of wonder. Gratitude is always in order.

And that's how I felt about the coffee that Ross made for us with painstaking precision and almost religious reverence. He uses an Aero-Press coffeemaker, and he brought his own select, freshly ground beans. Each cup was made individually and poured with care, as if it might somehow be ruined by being rushed into the cup. We were staying in a nice little Airbnb in the tiny town of Spruce Creek, and the cup I chose from the cabinet had a drawing of a happy puppy on it with the words, "An Ordinary Dog Can Make Life Extraordinary." So true.

It seemed appropriate as we walked down the dirt road to the confluence of Spruce Creek and the Little Juniata with our coffees in hand. While we walked, I listened to the birds and watched as the river grew closer with each descending step. And then I stood there with my friend, looking over the railing of the bridge as the river flowed beneath us and within my waking dreams. I was saying "farewell" to this river, and I meant it.

Why do we always have to part with the people and places we've come to love? (Or watch silently, as they part from us?) Just a single night prior I was standing in this river's cold, quick waters while the mayflies lifted into the skies all around us, as we caught and returned more lovely trout than I would care to count—if I was the kind of man who cared to keep count. I'm not.

It was hard for me to turn away from the Little Juniata, knowing I might never see her again. Love is like that—bitter and sweet all at once. But as we loaded up the SUV and headed out toward Penns Creek, I knew that I was about to meet a new best friend, and knowledge like that takes the edge off the uncertainty of farewell moments. And besides, as wonderful as that first cup of coffee was, we needed breakfast.

We drove through more of the bucolic countryside of central Penn's Woods while listening to the melancholy Irish music of Dermot Kennedy. His lyrics of love lost and longing resonated with me and reflected the emotional and artistic depths of my friend Ross. We have this in common. My Norman Irish poet ancestors called to me through the sadness of the songs and the joy of having loved people and places so truly that our connection lingers long after they have vanished from our lives.

As a visiting Texan I am always overwhelmed at the verdant greenness of the eastern third of our country. In my western homeland the dry shades of gray and green are only offset by the slightly brighter green of ash juniper and the myriad of colors brought on by spring wildflowers, ever so briefly in bloom. But here a vibrant, almost shocking green dominated the landscape, interrupted only occasionally by the hand of humanity. Here beyond the asphalt roads were the Amish farms of white-painted houses and barns. Future family dinners grazed around each farmhouse, as seemingly unconcerned with their eventual fate as the average American is about the unfolding of our broken and burning world. Ignorance is only bliss for a little while. Eventually, the inaction of every sheep leads to a future of lamb chops and woolen coats.

We drove into a former coal town turned college town and pulled up next to a coffeehouse called The Brew–Coffee & Taphouse. When we walked in we were greeted by two friendly young women behind the counter, and their genuinely joyful demeanor added something special to the morning. It's as much a fact as gravity that our body's limbic system reacts to the behavior and emotions of those around us. Surround yourself with "victims" and you begin to feel like a victim. Surround yourself with angry people and you are in danger of becoming one and seeing the world in the same dark and threatening ways. Anger is a derivative of fear, and fear is often born in ignorance. The bottom line is that we become the company we choose to keep, in person or via social media. And it is a choice, just as inaction is a choice to be subject to the whims of others. Ask any plate of lamb chops. Sometimes we need to jump the fence.

The coffee and breakfast sandwiches were heavenly, and the conversation was hopeful. Ross and I talked about our destination of upper Penns Creek and what might unfold during the day ahead. He was excited about sharing another of his favorite home waters with me, and after our stellar evening in the Little Juniata, I was just as excited.

Penns Creek actually feels like a river with its wide stretches and deep holes. It begins just north of the village of Spring Mills where it emerges from a privately owned limestone cave. Penns Creek is one of the two longest limestone waters in the Keystone State—the other being the Little Juniata. And it's known for both its prolific hatches of mayflies, stoneflies, caddis flies, and midges, as well as its potentially selective trout.

In Dwight Landis's guidebook *Trout Streams of Pennsylvania: An Anglers Guide*, he writes, "Penns Creek has a reputation for being an unpredictable stream. There are days when the fishing is very good, and there are days when your best efforts will produce little or no success. Even the highly skilled Penns Creek veterans, who have fished the stream for years, get skunked sometimes." Reading this, I appreciated the comforting words that even the best anglers can struggle on this magnificent spring creek. As an angler, I'm nowhere near being "the best" at anything, so the pressure is off.

Driving to the river we passed through more Amish country, and I was taken by the irony of a muddy field full of Amish mules on one side of the road and the asphalt parking lot full of modern farm tractors at the store on the opposite side. Ross turned onto a winding farm road that was populated by Amish farms on one side of the road and modern farmhouses complete with electrical power and shining SUVs on the other side. And then we came to a sign that I soon discovered was a precursor to wonderful fishing in this part of Pennsylvania. It read: "No Outlet."

A red-eyed vireo was singing in the trees above us as we strung up our rods. The day prior on the Little Juniata, I had used the beautiful bamboo rod that suffered an untimely demise. So on this day, Ross lent me his Orvis H-3 5-weight. It was love at first cast. I was a lost man. The rod felt as if it were a part of me, and I wondered if it was a sin to covet thy neighbor's fly rod. If it is, I had some penance to do.

We walked a long way down a riverside path through forests that were much more forgiving and free of poison ivy than the woodlands surrounding the Little Juniata had been. Here the trees were tall hardwoods on our side of the river, with equally tall conifers on the opposite side. The undergrowth along Penns Creek was far less dense than along the "Little-J," and this made walking through the forest all the more pleasant.

Occasionally we came to areas of bog that had become muddy and rutted by the passage of anglers through the years. I tried my best to walk on the more stable part of the bog so as to not add my own footprints to the damage already created. As someone who sees the earth as a living being, I am careful about causing injury—as much as is possible. I'm fully aware that my mere presence causes some injury to the forest's floor or the fish's lips, but I try my best to treat both with respect. I never carelessly gouge the earth or "rip lips." Sometimes the latter leads to lost fish, but it never leads to regret. For me "the tug" is not "the drug." If I could release fish without catching them first, I would. I'm strange like that.

While walking the path, stopping here and there to recon the river and watching for likely pools or rises, we came upon two fellow anglers who we began to refer to as "the two guys and their dog." It was obvious that these two gentlemen were long-term fishing companions and that the dog was at home with them both. The pup was a handsome Norwegian elkhound named Popeye. The men did not offer their names, and one spoke to us in a casual friendly manner, while the other only spoke to the dog. The dog was egalitarian in his exuberance for friendship.

The two guys seemed nice enough, but I immediately knew that I preferred the happy dog who kept them company. Both men were nymph fishing due to the high and quick water, and the talker shared that up to that point in the morning, they hadn't enjoyed any luck in catching fish. We wished them well and moved on down the trail. I glanced back and waved to the dog. He wagged his curled tail as best he could.

Before long we arrived at a wide place in the river with a small oval island about two-thirds of the way across. The island was covered in knee-high grass and in the center was a Canada goose nest containing a single unhatched and abandoned egg. Not everyone makes it across this dimensional boundary—at least not on the first try.

There were a couple of beautiful long stretches of river here with semi-deep pools and good currents that slid past some natural structure of half-submerged trees, stones, and undercut grass-covered embankments. It felt perfect, and so we divided the river in half with Ross crossing over to the coniferous side and me fishing along the hardwood edges. Ross did the smart thing and went in with a double nymph rig while I did the stubborn thing and tried my luck with a dry fly.

The day prior on the Little Juniata, we certainly caught as many fish as we wished, but almost the entire day was nymph fishing and by now you already know "it's not my jam." So, today I decided to focus on fishing, not catching. Today I told myself that I'd pretend to be a "purist" and stick with my dry fly—come what may. That's what I told myself.

I watched as Ross carefully waded across the big, wide stream, just downcurrent from my position. He was easy to see since he was wearing a neon-blue casting shirt, while I was once again perhaps foolishly wearing my camouflage shirt and hat that I had gotten into the habit of wearing long ago while immersing myself in small-stream stealth fishing. The day prior Ross had been dressed in bright yellow. He looked quite dapper while I was busy doing my best imitation of a willow tree. Did I mention that both Ross and Dennis caught more fish than me? Well, they did.

I began casting and getting some wonderful long drifts through what seemed like perfect water. That rod vanished in my hands and became a part of my amphibious soul. I still can't understand exactly how or why. I've cast high-end rods before and more often than not, I preferred my relatively inexpensive glass rods. Fly rods are like bottles of wine. Just because something is "fancy" doesn't make it right—for me.

Most people don't know that the Texas Hill Country is wine and vineyard country. Some of our vineyards produce expensive and presumably fine wines and I've tried many of them, but the one I drink everyday costs seven dollars a bottle, and I prefer it. With that said, I once had a bottle of expensive Italian "Super Tuscan" that seemed like an extension of my soul, and now I was casting a rod that felt much the same in my hands. But even with that marriage of perfect water and close to perfect presentations, I remained fishless. And that was about the time I heard Ross make a whooping sound and saw his rod bent over as he landed another nymph-caught fish. My dry fly looked so lonely, just floating there, uneaten, unloved, underappreciated. I kept the faith and cast again.

Ross and I met up on his side of the river, reeled in, and climbed up the shoulder-high soil and stone embankment and into the piney woods. Here, the riverside trail was quite open and made for easy walking. The path was largely unobstructed, with only the occasional mountain laurel leaning over the trailside, and the fallen needles of pine and hemlock softened our footsteps. Thousands of tiny blue and white flowers covered the forest floor wherever the sunlight managed to shine between the trees. And the trees shaded and cooled the water while providing shelter for the trout.

The occasional sulphur mayfly rose from the river, only to eventually cling to the trees and then fall back into the water—home. A mink fished for his breakfast beside us while keeping a wary eye on the two suspicious beings who were standing in the river and waving long sticks in the air. And it was about then that the mother and father geese swam by us with the three goslings that did manage to hatch closely in tow. Life goes on.

There is nothing we can do for the souls we leave behind. All we can do is keep casting into the sky and drifting on the water. All we can do is continue walking our own path, wherever it may lead us.

I never lose sight of the understanding that I am walking on soil—not dirt. Dirt only exists where humans have killed everything that's supposed to be alive in the soil. This beautiful blue and green planet is not covered in dirt, it's alive with a skin made of soil. And soil is a living, thriving community of plants, animals, fungi, and microbes that feed each other and feed on each other. Even the stones create the conditions for life. Just ask any caddis fly nymph in the river, or salamander in the soil.

In his Pulitzer Prize–nominated book *The Forest Unseen: A Year's Watch in Nature*, David Haskell tells the story of his year of intense observation of a section of Appalachian forest floor that fit inside the hula hoop that he had placed there, thereby creating the artificial boundaries of a boundless world. During that year, he witnessed and recorded the drama of life unfolding in a rounded meter of the forest—life and death and new life. Seasons brought changes and changes allowed life to persist through each season. And in the end, there was no end—yet. If we simply choose to pay attention, we will know the miracles that are occurring all around us.

There is a type of walking meditation that I practice on numerous occasions. Whenever I walk along a trail, I focus my attention on my connection to the earth. I *feel* my footsteps and the weight of me upon the earth, which is the result of the weight of the earth upon my footsteps. Gravity meets gratitude. I am held in each moment, step by step. I become more aware of my connection to this beautiful planet we've been given. While I do this, I practice mindfulness through my eyes by clearing my mind of "thoughts" and simply *seeing* what's all around me. I pay attention to every blade of grass, every leaf, and every stone. And then I meditate with my ears. I *hear* every cricket, birdsong, the step of a deer in the leaf litter, the slither of a garter snake in the soft morning grass. I miss nothing. I am present and aware. I am not wrapped up in my artificial thoughts. I am completely alive—feeling, hearing, and seeing the world around me. Try this the next time you walk into a river. Be the river, not the fish.

We came to a place in the river where some massive stones jutted from the water and the earth. We sat on the earth stones and ate our lunch, and we stood on the river stones to cast our lines. After lunch, Ross asked me to help him record a video that would end up on the website for *Fly Fisherman* magazine. I was all too happy to do so. As I held his iPhone and acted as cameraman, Ross played the roles of director and talent, which is a good thing, since I have no such talent. During the video he gave some handy tips about keeping your dry fly dry—and I actually learned a thing or two. For example, you should use gel floatant only *before* the fly gets wet, because its purpose is to fill the spaces between the fibers so that water doesn't get into those air pockets and cause the fly to sink. Once the fly is soaked, it is too late. And then there was the part about using a paper towel to squeeze the water out of your saturated fly before dusting it in desiccant powder. It was all valuable information filmed against a beautiful background. I enjoyed my tiny role in the process, but it was time to get back to fishing.

This is about the time that I temporarily gave in and snipped off my dry fly and tied on a nymph rig. The river here was deep and fast, so I worked my way out onto the treacherous rocks and drifted the nymph and bobber contraption through the runs and pools with a guarded sense of hope. Ross walked downriver and out of sight, although I knew which deep and rocky pools he was heading toward.

When I got bored nymphing, I cut it all off again and tied on a wet fly simply because I felt like it. There was no magic or mumbo jumbo to it, and certainly no deeper understanding of what the trout might want. I simply wanted to swing a wet fly for a while. So I walked back upriver until I ran into the two guys with the dog again. I made sure to give them plenty of space before I waded into the river at the head of a nice long pool and began swinging the wet fly in a peaceful, methodical manner. I did this for quite some time while working my way down the pool and across the river to another nice run, and in the end . . . I caught nothing. I would later learn that Ross picked up two more fish while patiently and expertly nymphing those fast, deep pools that I had abandoned. Experience and expertise beats exasperation every time.

Clint Eastwood's character Dirty Harry used to say, "A man has to know his limitations." It wasn't meant as a compliment when he said it, but it remains true, to a point. I'd rather restate it as, "We all have to run our own race." Ross had caught three fish on a slow day with his nymph rig, but the two guys with the dog reported that they still had not caught a single fish. So I decided that it was more important for me to enjoy this beautiful day on the water, doing what I love to do, than drag artificial bugs over the gravel in the hope of upping my odds.

I cut off the wet fly and tied on a caddis dry fly, just because I'd seen a few flitting around and even had one land on my arm. Then I walked the trail back upstream toward the beautiful wide place in the river where we began our day. I wanted to give that spot another chance, but that was where I met "the slobs."

There were three of them sitting together on the bank of the river right where I wanted to fish. They were having lunch and sipping a beer as I walked up to their temporary encampment. They smiled and waved at my approach and immediately offered me one of their ice-cold beers from their cooler. That's when one of them said, "I know we look like a bunch of slobs but we're really not." And then the guy next to him laughed and said, "Well, we kinda are." We all laughed.

I learned that they were three friends from Ohio who make this pilgrimage together every year, riding their Harley Davidson motorcycles to the Poe Paddy State Park campground. This was their one chance for a few days away from the responsibilities of work and family life, and they all admitted having understanding wives who sent them on their way because it was "good" for them. And as it turned out, they had decided to lay claim to that section of river and wait there until the expected evening sulphur hatch. Fair enough. It sounded like a good plan, and I wished them well as I crossed the river and began walking upstream to find Ross, who had leap-frogged past me while I was swinging the wet fly.

I really liked those guys, at least as much as I liked the curly-tailed dog and even more than the dog's companions. And quite contrary to their joking commentary about themselves, in no way were they slobs. These were just family men on sabbatical. I'm cheering for them.

I met up with Ross on the trail, and we decided to go back to the parking area and take a little break to enjoy a cold brew and eat some more food that I shouldn't be eating. I had surrendered to the current reality that meat and cheese sandwiches and potato chips was what there was, and if it killed me, I'd die happy. And I was happy even if my cardiovascular system was not.

On the way up to the parking area and our picnic dinner, we came across "National Park Guy." We called him that because when we met him he was wearing a floppy hat with an embroidered National Park Service patch and a T-shirt with the same insignia on it, and during our brief trailside conversation he told us about his journeys around the country visiting various national parks. He seemed like a nice man. He was wearing an old-style fishing vest and carrying a whippy-looking bamboo rod. We saw him again as we walked back down the river looking for our new place to wait for the expected evening hatch. He had joined up with the two guys with the dog, and it still looked like none of them were catching anything. The dog seemed bored.

And then we found it: the magical place we had been searching for and had evidently walked past several times, but in the evening half-light it seemed somehow more than it had earlier in the blinding brightness of midday. That's how discovery is sometimes. We need to keep looking with open eyes, mind, and heart until the universe reveals itself to us—in its own time.

For me, joining a river or mingling with an ocean is a spiritual act, as is the act of discovery. For me, discovery is spiritual in that it is like finding what you already knew but somehow have forgotten. I don't have any faith in human-created religions. For me, these are stories of mythology that are intended to control us and comfort us, all at once. I have no religion to preach. My faith is in the act of discovery, or what might be better described as "rediscovery." My faith is not rooted in a delusion of remaining "me," but rather in a realization that at a quantum level "I" am simply part of "we." And "we" are not simply humans. We are a part of every living thing—including the earth itself.

This place felt like that as we waded into the river, and I found my spot within its moving waters. Ross waded a bit farther upstream so that we could be in solitude—together. Already, in just a few days on the water, I had discovered that Ross and I have much in common, including our mutual love of *Star Trek*. (I bet you didn't see that coming.)

While we stood there waiting for the sulphur hatch, we began chatting about how we both grew up watching the original *Star Trek* and how our love for the show had ebbed and flowed through the various new versions of the original story. We chatted about how the original show was so far ahead of its time, not just in the idea of interstellar travel and exploration but in the visualization of an egalitarian society where humanoids lived, loved, and worked together without regard to purple skin, pointed ears, or differing points of view. Ross preferred the original Jim Kirk leadership style, while I confessed to being a Jean-Luc Picard leadership devotee. And as we stood there quietly discussing the potential impact that the green dancing women of Orion had on our adolescent former selves, the sunlight began to fade into soft golden tones and the sulphurs began to rise up off the water faster than a tribble can replicate.

Quickly, the previously unbroken surface of the water became punctuated by rising trout. Our hopes rose with them.

Ross and I began casting toward the many trout that had seemed so completely absent just moments before but now surrounded us at every turn. I threaded my back cast between the trees and landed the sulphur dry-fly imitation ever so softly on the water and exactly where I wanted it in the feeding lane. And I remember thinking that I was not capable of such perfection and that it had to be the rod—not the operator. But sometimes every man simply needs to meet his soul mate before he can be fully in sync, or at least that was the mythology I created in my mind as one cast after another landed right where it needed to be, and miraculously, astoundingly, almost inconceivably, I caught nothing, and neither did Ross. Something was amiss, but what?

As the evening light dimmed, the hatch turned into a spinner fall of truly epic proportions. Dying mayflies were tumbling toward the river and slipping away with the currents unless they were sipped away by the rising trout. Ross had a few mayfly spinner imitations that looked darn close to the naturals, so we each tied one on and began casting to the same rising fish that had ignored our dry flies. The trout were rising all around us, including within arm's reach. We casted and drifted until Ross looked at me and said, "I'm thinking we try to make it to town for a real dinner." I looked up between the trees to see the first stars twinkling in the night sky. I tried to look out across the river, where I could still hear but no longer see the rising trout. "Good food, live music, and a cold beer, you say?" He didn't have to suggest it twice. After all, a man has to know his limitations—it's important to run our own race.

Chapter Fourteen

Floating Lower Penns Creek

Stay close to any sounds that make you glad you are alive.

—Hafiz

It was just one day prior that my friend and tribal brother Ross Purnell and I had shared a day wading and wandering together along the forested edges of Upper Penns Creek. We fished from the glow of first light until the arrival of the soft, gentle, half-lit evening. When that day receded, mayflies tumbled from the heavens, weary of flight yet content in giving their last full measure. Together we had come full circle, returning to the home waters of our ancestors. As a young man, my father once cast his line across these Keystone State waters.

It was a day that I will never forget. It contained all the most important elements of a truly magical and meaningful fishing trip. Birds sang in the bright green trees as conversations and moments of silence and solitude were gratefully shared. The river ran quick, cold, and clean around my legs, carrying my cares away within its persistent currents. It was a perfect day. Did I mention that I caught nothing? Ross caught a few. It doesn't matter. I loved every fishless moment.

And now it was a new day as we gathered together to float down the same river—transformed by its accumulation of warmer waters. Nature is full of paradox and parables that act to teach us the power of perspective and remind us that change is our only constant. Even gravity might someday vanish, allowing whatever remains of this wounded planet to float aimlessly like mayfly spinners on a global twilight riffle. Until that day we still have the opportunity to become our better selves and choose which harbor is the right harbor. Drift boats don't really drift unless we surrender our ability to take up the oars and choose our response to the pull of the currents and push of the winds. This is true in the life of our rivers and the rivers of our lives.

This day felt more like a celebration than a contemplative moment of solitude on the water. There needs to be room for both celebration and contemplation in our precious and finite lives. We humans are social animals, and it's been demonstrated with data that those who enjoy healthy social relationships live longer and with better quality of life. I don't know about you, but I'm "all in" for that life and those outcomes! As I shared with my cardiologist, "I'm not afraid of being dead. I was dead for billions of years and it didn't bother me a bit. I'm just not ready to leave. I have so much yet to live for and share."

On the way to the river, Ross and I traveled through more Amish country, past the white-painted farmhouses and barns without power beyond what could be given by human hands or the forced labor of horses and mules. Black, boxy buggies with teams of horses in front and straw-hatted, long-bearded, black-and-white-clothed men driving the teams crossed our path from time to time. Usually the driver stared forward without a blink or a glance, as if our existence was unseeable and unfathomable. On occasion we'd manage to make eye contact, and a friendly wave showed recognition of each other's humanity—from one worldview to the other and back. We need more of that, I suspect. Not merely tolerance, but rather actual acceptance and natural empathy. There are so many fences to mend or, more accurately, to tear down.

When we arrived at the Earlystown Diner with its gravel parking lot and unassuming facade, Ross was eager to introduce me to this regional gathering place, and I soon understood why. Walking in, I was struck by the decor almost as much as the mingling of generations and the immediate kindness displayed by every new face I encountered. An old man wearing coveralls and work boots and pulling an oxygen tank behind him insisted that he hold the door for us. He had a bright smile and a twinkle in his shockingly blue eyes, and it was immediately apparent that no amount of medical gear on wheels was going to slow him down. Walking next to him was a woman I took to be his wife. She too flashed a soft smile as she looked at her husband and then back toward us. These are the people who become my heroes in an instant. No amount of hardship will dampen their gratitude and joy. He reminded me that even with my wounded lungs and broken heart, I can and must keep living and giving and opening the next door for others and myself. Every day is an opportunity and a gift.

The gray-and-blue-painted wooden walls were decorated with paintings of farms with vintage pickup trucks—each with blankets of flowers growing from their beds. Wooden hand-painted signs declared, "Happiness Homemade," "Good Food Served Here" and "Today's Menu Is Take It or Leave It!" Bright orange bench seats were paired with 1960s-era Formica-topped tables, and in the back was a counter from a similar era. I could almost hear the echoes of Hank Williams singing from somewhere far away.

Seated at the counter and surrounding tables was a cadre of jovial locals who all seemed happy to see each other, which caused me to wish that, just for this morning, I could be one of them. But I wasn't, so Ross and I took a seat up front by the window where the morning light slanted in so that the Americana painting we were inhabiting seemed to be simply perfect in both light and composition. And that's when we met our server. I will never forget her—not for how she looked, but rather for her genuine ability to reach across the generations and show true affection to both the elderly gentleman at the counter and the young couple sitting together at what must have been "their table."

I don't recall her name, but it doesn't matter because it was her gentle nature that identified her more than anything else. She wore a bright green T-shirt, bright blue-painted fingernails, a swirling kaleidoscope of tattoos, and an obvious affection and kindness for every regular customer from rambunctious child to ancient farmer—all laughing and living together in this moment, as it should be. And she was kind to me and Ross, but I knew there was a natural difference between the old and young locals who regularly filled these tables and a couple of anglers who were just passing through. She knew this, too, even better than I. Some understandings need no explanation.

Throughout my visit to central Pennsylvania, I found it nearly impossible to find any food that fit my Nutritarian diet. Even when I could find greens and beans, they came dripping in butter, cheese, and salt—"no substitutions." And wherever I found fruits, berries, and nuts, they either came baked in a pie or salted and smoked. So I surrendered to the regional all-American cuisine—with a few solid boundaries. One such boundary was "scrapple."

And this brings me back to the "standard American diet." I've discovered that almost everywhere I traveled, my heart-healthy habits would have to relax and hope for the best, or I'd starve to death because there would be nothing to eat. In this case, Ross wanted me to try the scrapple, which is a regional concoction made of boiled pork scraps (everything but the oink!) that are mixed with buckwheat or another grain, then the entire cardio-catastrophe is formed into a "cake" that is fried or deep-fried, just in case the pork suet and salt wasn't enough to stop your heart.

Now a younger me absolutely would have tried it, but that younger me didn't know about the coronary heart disease in his future and all the choices he made to contribute to it. There is a fine line between brave and stupid, though, so this happy traveler took a pass on the scrapple. Still, I needed to eat, so I allowed myself the extravagance of eggs, dry toast, and fried potatoes—none of which was Nutritarian, but all of which was absolutely delicious, almost "to die for."

After breakfast we headed toward the river while listening to Ross's music selection of the day—Metallica. To add to the color of the morning as the cholesterol and sodium chloride coursed through my aging arteries at the speed of sludge, I learned that Ross had once been in the business of managing local events and had an opportunity to work with the band Nirvana and Kurt Cobain. We chatted about his once-in-a-lifetime experience with Nirvana and the Red Hot Chili Peppers and about how we've both lived many lives in this single lifetime. It was a memorable drive past the Amish barns and forested winding roads where pileated woodpeckers darted between trees and screech owls called over the haunting sounds of Nirvana's Kurt Cobain singing his acoustic version of David Bowie's song, "The Man Who Sold the World." In a way, we've all been that man.

Everywhere I go and everything I do seems to have a soundtrack. This one suited me just fine as we drove down the road between one community that for its own reasons is trying to hold the world at bay and another that seems to hold nothing back—but reason. One world refusing to be "public" while the other fails to keep anything "private." And as we drove toward the river, I wondered how much of our world and our future we've "sold." Nature isn't a capitalist regime. It seems more like an egalitarian barter system to me—carbon dioxide for oxygen and oxygen for carbon dioxide—"winner takes nothing." Every child with a slurpy and a straw knows that if you keep sipping, eventually you're left with nothing but an empty cup.

When we arrived at the river, we met up with our fellow adventurers for the day, which included Ross's son, Carson, Jay Nichols, and Ben Annibali. This was my first time meeting each of them and, luckily for me, we became fast friends. To say that Carson Purnell is extremely intelligent and interesting is a drastic understatement. He is also a hell of a nice guy. Jay was described to me as a man who is "amazingly good at everything," but whose modest and warmhearted demeanor belies that brilliance. I found him to be exactly as advertised. And Ben is one of those young men who immediately give a much older man like me a bright light of hope, just when I'm desperately seeking such youthful illumination in these all too often seemingly dimly lit times. We were five guys and two rafts on one beautiful river, sharing a gorgeous day, which just goes to show you that there is more to a happy heart than just eating beans and greens. Part of being healthy is enjoying good friendships and great moments that are shared in Nature. The mind is as much a part of healthy living as the body—perhaps even more so.

The lower portion of Penns Creek is bass water. And as much as I enjoy the ballet of catching trout, I am a bass angler at heart. Being a son of the South, I grew up tossing spinners at largemouth bass and, later in life, casting a fly rod to Guadalupe bass in my Texan home waters. After two days of daintily dunking nymphs and delicately presenting dry flies to softly rising trout, it was going to feel good to enter the equivalent of a barroom brawl with a truly American fish—the smallmouth bass. I couldn't wait to chuck a streamer against an undercut embankment and watch the water explode! And it didn't take long for my dreams to come true.

I was slated to begin in the bow of the boat that Jay was navigating, and I might as well say that I remained there for the duration of the journey. Ben manned the back of the boat, and as soon as we got into good water with a little cover under some massive trees, he began to connect and landed three bass before I could manage one. But then it was my turn, and the surface of the cove transformed into the kind of splashing, slashing, swirling battleground a bass man like me lives for and loves. We'd only been on the river for a few minutes and already the action was furious.

I looked upriver where Ross was rowing his boat with Carson fishing, and I could see that Carson was into fish as well—although silently. I was to learn that Carson never whoops or hollers when he catches a fish but rather maintains his constant quiet, calm demeanor as if nothing had ever happened. I enjoy seeing the different ways we all react and interact while fishing. Carson was having fun—quietly. I liked that.

Ben and I were much more traditional in our exuberance. I laughed when I blew it on a hookset—and I laughed often. Ben and Jay traded commentary back and forth as we each took turns catching and sometimes failing to catch and land fish. Most of the bass we were landing were of moderate size, but a few were pretty hefty, including a big female that I hooked into in a back eddy where the water was so clear that I could watch everything unfold even within its depth and turbulence. It took me awhile to land that fat fishy as she bore down to the bottom of the hole, testing my tippet and temperament. But she was well worth the wait, and I was grateful that I managed to stay focused and composed until we got her boatside. (It was a nice change from my previous performance of overpowering and breaking off a perfectly catchable fish!) She was a hefty hunk of bass flesh, and although I tend to shy away from grip-and-grin photos, I was all too happy to hold her briefly as Jay snapped a photo. I'm a conservationist, not a saint.

I was casting a 7-weight Orvis H-3 and tossing a big tan articulated Gamechanger. I had fallen in love with Ross's 5-weight H-3 on upper Penns Creek the day prior, so he lent me another one and it did not disappoint! I don't know why, but that rod feels like an extension of my immortal soul! As for the Gamechanger, I had to change my game from my usual Texan practice of slinging the streamer up against the bank and straight-stripping it back toward the boat. The trick here was to toss the fly, let it sit and sink a moment, and then mend the line repeatedly so that the articulated fuzzy contraption wiggled and wagged in the water like a wounded forage fish. All too often the strike came as soon as the fly hit the water. Bass aren't shy creatures nor are they persnickety. They'll eat just about anything they can get in their maws, from a dragonfly to a duckling.

Ben was throwing a yellow Boogle Bug popper and getting hit after hit on top water, while my streamer was more of a here-and-there proposition. I admit casting an envious eye toward Ben's topwater popper, but Jay advised me to give the Gamechanger a bit more time to change the game—and I did. As fish after fish seemed to take great offense at the mere existence of Ben's Boogle Bug in their part of the river, I practiced my Zen patience by repeatedly mending the line until eventually a bass would strike if for no other reason than sheer pity for the old dude bent over the bow of the boat looking "hopeful." Fish were being caught in a rather steady stream of retrieval and return, and I was amazed at the volume of bass biomass in this beautiful Pennsylvania river. This Texan was in heaven. I suspect that heaven and hell are what we make of them, here or wherever "there" may be.

After traveling quite a few winding miles down the river together, we brought the boats into a cove and anchored for lunch. Ben had prepared and packed our lunch and Ross brought the beer; as it turned out, the right men were chosen for each job. Lunch consisted of a vacuum-packed epicurean delight that included roasted chicken with sun-dried tomato and goat cheese sandwiches on fresh ciabatta bread. (That's right, Ben took the time to vacuum-pack the sandwiches.) And just as I was thinking that all this meal required was some white wine, Ross handed out something that made me forget all about chardonnay—Mango Cart beer. It was love at first sip.

As I had learned earlier on this trip, Ross Purnell is not a beer drinker; he is a beer connoisseur with a finely discriminating palate. He had already introduced me to a few local treasures, and now I was forever grateful for the gift he had given me in the form of a mango-colored can that read "Golden Road Brewing, Mango Cart, Mango Wheat Ale" on the front and had the Surgeon General's health warning on the back. Meat is not part of my heart-healthy diet, but the chicken was delicious, and as for the beer, I'm not pregnant or driving a car so I decided to simply savor the moment and relax. Yah mon. It's not only beans that are good for your heart.

While we were eating lunch I enjoyed listening to Ross and Jay chatting about the art and business of publishing for the fly-fishing community, from "how-to" and guidebooks to outdoor literature. They chatted about frog flies and damselfly flies and places they'd been and people they'd met while casting a line through distance and time. And then we chatted about my passion for writing and how if someday I can no longer write, I'll probably be ready to cross the river and rest beneath the shade of the trees. For me, writing is living. It's how my soul breathes.

After lunch we switched things up, and Ross joined me and Jay while Carson and Ben went off on their own adventures as we all floated downriver together—yet apart. I had caught fish with some regularity during the morning, but it was becoming apparent that Ben and his Boogle Bug were just the right matchup for unlocking any mystery to these fish's desires. So Zen patience be damned, I switched to top water!

Jay was selfless in his desire to continue to row while Ross and I took turns casting up against the bank and under the limbs of half-fallen trees. And it was about then that I proved the point that "everything is bigger in Texas," including when we screw things up. Somehow I managed to cast my fly perfectly into the arms of an overhanging tree so that if it had been an Olympic sport, I would have won the gold medal. Jay was patient as I untangled my triple flip, double French twist, back-rolling half hitch from the tree. No Mango Cart for me until I redeemed myself.

I guess this is as good a time as any to share a new tradition I began on this float: Every time I caught a fish I'd yell, "Mango Time!" and we'd all take a drink. We were in no danger of becoming inebriated. Still, it added to the fun-loving and lighthearted atmosphere of our day as even a blind pig finds a truffle and Ross had an ample supply of libations for the duration. (Yes, I also drank water.)

There wasn't any part of the river that didn't seem to hold fish, but some areas were simply schooling with good-size bass that seemed to be eager to smash anything that landed on the water. At one point I tossed the fly out into what I thought was an unlikely trench in the center of the river, and even there a nice bass rose up and smashed the fly. I worked it into the boat, wet my hands, and lifted him carefully so that I could quickly return him to the water. But he had taken the fly in deep, and I asked Jay if I could borrow his forceps, and that's when more laughter broke out from bow to stern. They were the tiniest set of forceps I've ever seen. Had I tied them to the end of my tippet I think the bass could have eaten them. I started laughing and said, "Jay, your forceps are so damn cute! Where's their momma?" After trying to convince me that "size doesn't matter," Jay retrieved his tiny tweezers, and we all had a good laugh. As I set the chunky bass free I noticed that he didn't seem the least bit amused by the encounter. He was a handsome but humorless fish.

Jay moved us into position along the left bank, where if you were to dream of perfect smallmouth water, this would be it. There was a constant and steady current coursing through half-submerged deadfall mixed with rocky stretches and undercut embankments. Overhanging branches provided safety from above, while a mix of coves and protrusions made the entire area look like a gauntlet for any living creature small enough to fit in a bass's gullet. Ross had been casting and catching with a regular rhythm that bordered on monotonous, but now even time itself seemed to stop as something massive rose up from the depths and engulfed everything he was attached to except the boat. Ross played him perfectly, somehow managing to keep him away from the submerged structure while also not breaking the tippet. At one point this big-bellied bass tried to jump, but he could only manage to get his body about a third of the way out of the water before he submerged like a breaching whale. Jay started to laugh and mimicked in a voice that I suppose was "bass-like": "Oh, I tried to jump but I'm just too fat!"

Like me, Ross doesn't take many grip-and-grin photos, but this fish turned out to be his "personal best" smallmouth to date, and so Jay snapped off a few shots before he was lowered back into the river. We all watched as he sank into the river's deep darkness like a gangster with concrete boots. I think we all felt a mixture of childlike joy and mature reverence for this fish. I know we all felt grateful. And I, for one, considered asking Jay if he had a harpoon on board as we watched Ross fight that mighty beast, but I didn't ask because I felt sure after seeing those forceps that if he had one, it would be too small.

The sun was behind the trees by the time we came to our takeout, and as is my habit, I stopped fishing for a while and just sat in the bow trying my best to absorb every aspect of this place and time. I took notice of the sound of the birds and the rhythmic rolling, swirling sounds of Jay's oars pushing us ever forward, like the march of time and inevitability of entropy. I listened to the murmurs and laugher of people who were at a riverside campsite called "Little Mexico," and I grew thoughtful as I saw them all living and loving in a community of tents and trailers, complete with coolers and cookers and all the comforts of home.

And there was that word again—"Home." I still don't know what it means to me. Every place I once thought of as Home has changed beyond recognition. It's like having an old girlfriend who ended up strung out on drugs and ruining her life. You may still love her, but in truth, you love who she once was before the fall. Drained marshlands and dried-up rivers leave only memories and regret. And that, my friends, doesn't feel like Home.

Recently, after I received my diagnoses of heart disease, "evolving asthma," and borderline severe sleep apnea, I asked my wife a pretty deep question. She had just gone through her own health scare at the same time we lost her wonderful mother to COVID-19, and I was still pondering the loss of my father, six years after his passing. I asked her, "If my heart gave out and I suddenly died, what would you do next?" We are both in our early sixties, and these questions seem more poignant with each passing day—not in a morbid way, but rather in an enlightening way.

She understood my question, and after careful consideration and an honest admission that she wasn't exactly sure, she said, "I'd sell the house, because without you here there is nothing else that would make me want to stay." And then she said, "Perhaps for a while, I'd get a short-term rental flat somewhere in Europe and travel a bit until I figured things out." This made sense to me since our daughter lives in the United Kingdom and my wife was born and raised in Scotland.

But the key takeaway was that she'd sell our house and travel a bit, without any specific timeline or goal other than allowing herself the time and space to figure out what comes next. It's a good plan, and as I thought about the same question in reverse, I realized how closely we had come to similar conclusions. There is value in beginning with the end in mind.

Then she asked me, "What would you do if the situation was reversed?" I said, "I'd sell the house, buy an Airstream and a nice truck to pull it with, and travel around America following good weather and camping in one beautiful natural landscape after another until I too figured things out." After a pause I said, "And I'd write a book about it."

She understood, and I'd like you to understand as well. I don't write books and magazine essays for any reason other than the hope that by sharing my journey and whatever bits of wisdom life has taught me, I might do my part in helping others, and this planet. After all, my story is our story, and our story is the story of the earth. We are all connected, even as we pretend to be "independent."

When I wrote my first book, *Casting Forward*, my plan was to have no plan. I simply set out to fish the entire Texas Hill Country until the answers to "what comes next," came to me—naturally. And it worked. The fishing led to a sense of peaceful openness, and the peaceful openness led to acceptance and understanding. I was listening as I waded and wandered from river to riffle. When we listen, we learn.

And what my wife and I learned in that conversation is the same lesson I learned when I picked up Ross's Helios 5-weight and began casting it toward the rising trout. Home is a feeling. Home is a connection that comes naturally and cannot be manufactured, but must be accepted.

Our house is not our home; it's just a place that gives us shelter from the elements and as much privacy as any of us can manage in a world full of "cookies" and "AI."

I haven't figured it all out yet, and perhaps never will. But maybe home is something that travels with us like an Airstream trailer. It adapts to wherever we find ourselves. We can feel at home with another human soul—a dear friend whom we love. But then they evolve and change, and the time comes for them to move on. This is the way of things. This is how life unfolds. And we can feel at home with a place, until either the place or our feelings change, and we move on. I felt very much at home when I picked up that rod and seemed suddenly incapable of missing my mark. We just fit together. We were at home with each other, even if she belonged to another man. Love is like that—unexpected and unstoppable. Perhaps love and home are the same thing—ever changing and yet timeless and unconditional.

The next day I flew back to Texas. I went fishing in my Hill Country home waters, and it all felt wonderful, yet not quite complete. I couldn't stop looking at the photos and reading my notes from the week of fishing in Pennsylvania with Ross and the other members of my tribe. This has happened to me before, in Alaska and Montana—I seem to leave a part of myself in those places that my friends call home.

After a day of solitary fishing on the Guadalupe River, I returned to my house to find a package waiting for me at the door. It was Ross's Orvis Helios 3 5-weight and a note from my friend. He had gifted it to me along with the words: "When I saw you with that rod, it felt like I was watching a Jedi who finally found his own light saber. I knew you had to have it." I don't mind sharing that this Marine got a little choked up and emotional about that gift and those words. I was so deeply touched by this selfless act of friendship and kindness. As I held that rod case in my hand and read the words written to me by my faraway friend, all I could think of was how deeply this past week and this thoughtful act really hit "Home" with me. Life is beautiful, y'all.

Part V

Birthplace of American Fly Fishing—The Catskills and Northeastern Forest

Chapter Fifteen

Floating the Delaware, New York

No one saves us but ourselves. No one can, and no one may. We ourselves must walk the path.

—Buddha

We all come into each day as we are, and as we choose to be. I have learned over the decades of life and living to simply accept each moment, memory, friend, and friendship as they come to me with as much grace and gratitude as I can. No day is all we might hope for, and yet every day is more than we might ever imagine. And although I endeavor to live each moment with mindfully open eyes and a grateful heart, I must admit that it is in the transition times of morning and evening twilight that I find the most meaning, peace, and spiritual revelation. This was such an evening.

It was that soft, silent twilight time when day fades into night and sunlight becomes moonlight. It was that mystical, magical moment when it becomes so easy to quiet the mind and simply be in the moment and wait. The East Branch of the Delaware River was slick and silent when we walked up to it, with only a few small sulphurs and some slightly larger March browns coming off the water here and there. It was far too sporadic to call a hatch. It was more like a grin than a smile on the river's face, as if to indicate that joy does happen here—now and then—if we're willing to be patient.

So we stood there in the soft, wet grass, my friends Joshua Caldwell and Landon Brasseur and I, just watching the water and the air, wondering, and waiting for rising mayflies and trout. In time, both revealed themselves—infrequently. Just a few hours prior, I'd stepped off an airplane in Newark. We'd only have an hour or two before dark, but we came here feeling hopeful and mindful that every evening on the water has its own rewards without regard to the outcomes of connecting tippets to trout. It was already the perfect end to another day we would never live again. We take for granted what mayflies cannot.

And so we stood there with our rods in hand and watched the refraction of the sun's photons in the earth's atmosphere bleed from golden yellow to crimson red and into a blend of royal purple and onyx black. The tree line was edged in gold as if to remind us that even in the dark of night, the forest lives on. It is often good to simply wait and see what unfolds. It is always important to remember that all of "time" is now.

Josh Caldwell is my friend, but he is also a talented film director, producer, and writer who directed the movie *Mending the Line*, which recently premiered here in the Catskill Mountains at the Woodstock Film Festival. He and his wife, Danielle, live and raise their beautiful children here in a home that is surrounded by forests and filled with love. Landon and his family also call the Catskills home. He is an extremely capable fishing guide, and although I was fishing with him as a friend, if I needed a guide in these waters I would reach out to him without a moment of hesitation. And that's not just because he is a knowledgeable fishing guide, but also because he's a nice guy and fun to be around. I've gotten to that place in life where I won't fish with anyone who doesn't add to—not subtract from—the experience, no matter how talented they may be as anglers.

Both Josh and Landon are able anglers, but they are also different in their approach. I always enjoy the experience of adapting to and learning from my buddies' various styles. Everywhere I travel and cast into new waters I learn something about another way to present the fly, or find the fish, or twitch a popper. I'm a forever-novice, and I like that status. It keeps me feeling youthful—as any good forever-novice should. I will always have a beginner's mind. While many anglers learn a great deal about one place and way of fishing, I am learning a little about many places and ways of fishing and living. It keeps me acting bold yet humble—as any good Marine should.

Landon is laid-back but also has that quiet sense of focused attention that I see so often in my angling friends of the US Northeast. If I were to characterize Josh, I would use the word "driven." He takes his fishing seriously and wastes no time getting his rod strung up and waders on, walking quickly out to the river, ever mindful of the opening and closing of angling windows of opportunity. As for me, I'm not sure if it's my southern ways or simply what comes with age and mileage, but I like taking my time and chatting on the tailgate of the truck while watching the river for a while, not at all concerned about what I might be missing, because I'm not missing anything, really. I'm not lazy, but I might be laissez-faire in my approach to angling. I love to catch and release fish, but I won't let the catching of fish define my fishing.

The sulphur hatch was almost nonexistent, but we were seeing a few caddis coming off here and there. When people used to ask me what my "go-to" dry fly was, my answer was always a simple parachute Adams, but these days I'd have to say a caddis is my dry fly of first choice when I'm not sure which small brownish bit of fur or feather to choose. In part, this is because as I've traveled and fished across the United States I've found caddis almost everywhere, from Montana to Texas and Minnesota to Vermont. Another reason is that I can fish them in so many ways: drifting, skipping, bouncing, and dry-and-dropper combinations. And finally, I like that I can use them to imitate other insects like terrestrials. So we all tied on a caddis and Josh stealthily waded out toward the only rising trout we were seeing. Landon and I were happy to sit on the riverbank and watch Josh casting while the evening spread blood-orange colors across the treetop silhouettes. These are the times that seem to beat to the rhythm of peaceful hearts. We felt fortunate—together.

As the first crickets started to sing, a few more trout began to steadily rise in their feeding lanes. I waded far downstream of Josh while Landon positioned himself between us. I picked out a riser just below an overhanging dogwood, while Landon began watching a particularly large-looking trout that was consistently sipping the water some distance from where he stood. I could not have made that cast, but he did with ease and almost immediately connected with and landed a nice brownie. We all cheered for him, and once he had released the fish without ceremony or fanfare, we each turned back to the business at hand—namely, catching a fish of our own.

But as lovely as the evening was, it was also on the leading edge of a cold front, and I got the feeling that the fish knew something we hadn't discovered yet. Almost as quickly as the light faded and the temperature dropped, the trout sank sullenly with them. It was about then that I heard something crashing through the brush on the far bank of the river, and what I first mistook for the sound of a wandering bruin revealed itself to be the waddling and grunting figures of a pair of highly irritated beavers. One of the furry mammals began walking up the bank in my direction while grunting and the other plunged into the water and began swimming directly toward me. Based on my past encounters, this seemed like unusual behavior, but I was thrilled to see them, nonetheless. I always enjoy watching beavers.

Landon heard the grunting and splashing and asked, "Beavers?" to which I answered, "Yes, two of them." That's when he said, "Back out of the water! They might have pups around here somewhere and they might either tail slap you or bite you." I got out of the water. After all, it was their river; I was just visiting. And besides, the cold front was coming in, the fish were sinking down, and even the cedar waxwings had ceased to flutter over the water and found their roost for the night. Nature knows what we've forgotten.

We knew when we arrived that at best we'd have an hour or two of casting light, and there was an early morning ahead of us as we planned to drift the West Branch of the Delaware the next day. It would have been nice for Josh and me to catch something in that hour, but it didn't really matter. I was happy as we walked through the bending grass to the sound of crickets and the crisp feeling of the changes that were rolling over the horizon. What's not to love?

We shared tailgate beers under the stars, and then Josh and I enjoyed a conversation-filled drive back to my bed and breakfast where I'd rest up for the next full day's fishing. The anticipation of adventure is often at least as good as the act of adventuring. Ask any child on Christmas Eve.

My wife Alice had searched the internet while I was planning this trip and found the Inn at Stony Creek, and I'm so glad she did. Innkeepers Bill Signor and Joe Campone have done an amazing job of restoring this historic house and property that once served as a way station for the underground railroad—a safe haven for African Americans fleeing the horrors of slavery. The house, which was built in 1840, is now part of a bird sanctuary complete with a vibrant pond and flower-filled walking trails. It was also to be my home away from home, and no matter how late the hour when I walked through the door, I found Joe waiting to hand me a celebratory glass of wine. I slept peacefully that night and dreamed of rising fish and bending rod tips. It was a good night. I wondered how many dreams had been dreamed in that house.

It was fortunate that there was a deli open before sunrise each day, and Josh and I got into the pattern of stopping there before driving off into the dim light of dawn toward a hopeful day on the water. If you've read this far into my journey, you are already aware that I have been desperately trying to eat a heart-healthy diet and have found it nearly impossible to do so. Everywhere I have traveled, the "standard American diet" prevailed as if we had made obesity, heart disease, cancer, and premature death an act of personal patriotism. I'm no longer willing to carry that banner while at home, but while traveling I have adapted and made the best of my compromises. I will admit that I came to love my egg-and-cheese sandwiches each morning and told myself that I was only slightly sinning since I instructed the nice man at the deli to forgo the sausage and bacon.

What was different this time was my friend Josh's well-meant sharing of his belief in the high-protein and meat-rich diet that he and his family had adopted. I was not converted, but I understood that everything we do in life is an act of faith. I put my faith in veggies and beans and he puts his in meat and potatoes, and in the end we tap our cold cans of beer together and wish each other good health—and we mean every word of that wish. True friendship is like that. We embrace our insignificant, temporary differences knowing that what we share in common is significant and timeless.

The cold front had arrived, and this Texan had not come prepared for how cold that front became. I was wearing a base layer, a casting shirt, a fleece sweater, and a rain jacket. It wasn't nearly enough. While I slept the sky had turned ashen gray and the wind picked up as the clouds lowered. A soft freezing drizzle fell on my head as we loaded into Landon's drift boat and pushed off. I couldn't stop shivering. I've heard it said that there is no such thing as bad weather, only bad choices of clothing. I had forgotten how cold and wet an East Coast morning can get. I told myself that I'd choose better next time, if I didn't die of hypothermia this time.

The West Branch of the Delaware River was rolling high and quick. This part of the country had been experiencing the same extreme weather pattern changes as Alaska, Montana, Wyoming, and Pennsylvania. Unseasonably excessive rains here were the reflection of the unbearable heat and drought that was gripping much of Texas. It's important to understand that weather describes what's happening on a single day in a particular place. Climate describes the overall weather patterns in a historical and global context. Local climate changes may be seen in more frequent and severe and destructive storms, floods, droughts, tides, and wildfires. To put this in angling terms—weather might impact the fishing today, but climate impacts the fishing forever.

Natural climate change occurs over millennia, which allows plant and animals time to adapt. These current changes have occurred over a single industrialized century, which is far too rapid for natural adaptation to occur, and mass extinctions and species displacements will be the likely result. And here as everywhere I have traveled while writing this series of outdoor literary works, the same emotional climate has been expressed to me by people sharing their home waters with me: "Things are changing, and not for the better."

The Delaware River has been changed from its traditional smallmouth bass water to a "classic tailwater" by the building of dams and reservoirs that in turn provide a steady supply of fresh water to millions of people in New York City. Long ago Euro-Americans supplanted the native bass with non-native brown trout from Europe and rainbow trout from California that now populate these waters. Remnants of the original brook trout still hang on in the trickles of headwater pools and tumbling rocky streams. I'm grateful for those trickles and pools. As they sparkle in the dappled sunlight of the high forests, they make a sound that sings of hope.

For all that I see turning upside down in our dying and burning planet, I remain hopeful. If I were to ever allow myself to lose hope, I could no longer cast my line. I'd cease to find any reason to sit at my keyboard, writing love letters to the other members of my tribe and this beautiful, wounded planet. There is hope, and that hope resides in each of us. It matters not whether it's the heart that beats within my chest or the heartbeat of wild Nature in these ancient mountains, healing is possible—I hope.

Whenever I begin a float, hope floats with me. I imagine tight lines and laughter. I envision brightly colored leaping fish. I hear the dreamlike sounds of morning crickets and songbirds. I see the flashing of fireflies in the north and lightning bugs in the south, reminding me that there is no difference between glowing bugs or rowing blood brothers. It matters not if we originate from the American North or South, we travel together. Independence is a fairy tale we tell ourselves; interdependence is the true story of our mortal journey.

Even with the river running high and fast, the clarity was enough for me to see the healthy proliferation of aquatic vegetation waving like wildflowers in the wind. The roots of forest trees and meadow grasses held the riparian edges together and cradled the river in their loving embrace. There were good things happening here. A family of mergansers paddled by as wood ducks flew overhead and a bald eagle soared just downriver of our bow. After just a few casts my indicator submerged and I set the hook, bringing a deeply colored brown trout to hand. It all happened so quickly, yet also in a surreal timeless manner. I thanked him and set him free.

Landon was at the sticks of his craft and being a wonderful host as he pointed out likely places along the river and kept his head up even as the icy rain and sleet fell on us. I have lived in some cold, wet places. I have lived and traveled through the mountains of Scotland, Peru, Italy, Kenya, Montana, and Alaska. But in all the places I've traveled, I have never felt as cold as I felt on this day while drifting down this lovely river. I shivered uncontrollably. I felt frozen all the way down into every muscle, sinew, bone, and breath. I kept casting, hoping that each cast might take my mind off my temporary suffering and place it where it belonged—on how much I was in love with this beautiful river and my shared experience with friends. Love endures any hardship, and struggling well is a hallmark of healthy growth.

Even with the water flowing so freely, we came across areas of shallow gravel where the drift boat's keel might scrape if not for us all getting out and then walking the boat over the stony flats. I enjoyed that too. As much as I love drifting down a river, there is something special about standing in one and feeling the power of its currents against my legs. It causes me to slow down and look down into the river where the crayfish and sculpins wander. It causes me to be aware of my balance and to balance my awareness so that no bending grass, blooming wildflower, or singing warbler escapes me. And then we climb back into the boat, together, and our journey continues—just like life.

Tempo and texture make life interesting. The easy road is paved, but it contains no flowers, grasses, or buzzing bees. Give me a winding road of living soil any day. Give me a winding river and keep me away from canals and ditches. Predictable sameness makes me uneasy. Water, air, and lifetimes are best when moving and alive.

It had been a long and leisurely drift through verdant countryside that contained more forests than fences, and we kept rowing and drifting, casting and hoping, and eventually Landon noticed a single fish rising to something in a vast open bit of water. Soon, that fish became two, then three, then more than we needed to prompt the dropping of an anchor and a side-by-side effort to make a meaningful connection with one or more of these rising beauties. Both Josh and Landon are skilled anglers. Still, it was easy to tell that this was Landon's home water as he launched a leisurely but lengthy cast out toward a rising fish and quickly set the hook on a nice rainbow trout. I would come to see the same ease of understanding from Josh on the Farmington, and someday they will see this in me on either the Llano or Guadalupe Rivers of my beloved Texas Hill Country. Home water is more than a figment of our imagination; it's an extension of our being.

It wasn't long after that Josh's rod tip bent toward a living fish that leapt into the air and took out a bit of his line on a couple of nice runs before landing boatside. It was another beautiful rainbow, pink sided and speckled like a jewel as she vanished back into the depths but remained in my memory. Pretty fish and pretty girls have a way of lingering in an aging man's mind. We might not remember where we put the grocery list, but we never forget a spectacular rainbow and how she launched toward the sky before swimming back into the sacred waters that we were soon forced to leave behind.

The weather broke briefly, and so we anchored along a meadow-lined shore. After lunch we walked into the meadow and discussed the merits of fishing waders with zippers. Mine didn't have one. My raincoat also lacked hand-warming pockets, and it's possible that I harbored sinful feelings as I envied my friends' technologically superior status. I may have to rectify that situation. This became even more apparent as we pushed off for the final stage of our float and the temperature continued to drop as the breeze picked up and the freezing drizzle began to fall once again. I began daydreaming of hot coffee and warm clothes, which is not where my mind needed to be when there was fishing to be done and many miles yet to drift.

Every moment of this float was beautiful. If I'm honest, and I am, my Imperfect Texan Buddha nature did fail me once or twice. I know that I need to take everything as it comes and simply, as the Beatles sing, "Let It Be." But I will admit my human failings in that I wished we were connecting to a few more fish, even if only to help take my mind off how cold I was. And, as much as I loved this day, I will admit that the sight of the takeout point and the vehicles that awaited us, complete with heated air and access to warm coffee, was a welcome sight. The universe must have felt my weakness because that's when it happened—the "bad Buddha" test. Just as we were in sight of the boat ramp, Josh saw a fish rising within inches of the riverbank in an area with tricky currents and a winding bubble line. The challenge was accepted, and made to all of us as my breath danced in front of my face.

For Landon the challenge was to repeatedly and stealthily position the boat so that Josh could make a difficult presentation to what would prove to be an astute trout that had placed himself in the middle of a swirling mass of currents, submerged branches, and a grass-covered embankment. Josh's challenge was to make the cast, assess the fly selection, continue to be patient, rest the fish until he rose again, and then make a new cast while hoping to find exactly the right combination to unlock the door to this fish's heart. My challenge was to quietly cheer for my friend while trying to tuck my frozen hands under my armpits and not shiver so violently as to send shock waves through the water like a musky on the move. Did I mention that Josh is a deeply devoted and focused angler? Well, he is.

I'm not sure how many passes we made at that single rising fish. I am sure that Josh made one excellent cast after another, drifting the dry fly in a manner that seemed natural and worthy of success. And I am sure that Landon deftly and repeatedly repositioned the drift boat so that Josh could get cast after cast without disturbing the rising fish. And somehow the fish kept rising to everything and anything except Josh's fly. Finally, I'm sure that I failed as a "good Buddha" when I said, "This is beginning to look like your white whale, buddy." I instantly regretted those words because I knew I had broken the spell when Josh said, "I know, we're all freezing. Maybe I need to let this one go." To this day, I wish I'd kept my frozen lips shut. Josh earned that fish and in truth, I really wanted to cheer for him as he released it—but instead he reeled in, and we rowed to the ramp and got warm inside our waiting vehicles.

It was a tough day, but a good day. We had drifted many miles down a magnificent and legendary river. We cast our lines together and drifted our offerings into that aquatic universe hopefully, grateful for every moment and every mile, and we each caught exactly one trout. The fishing was wonderful, though the catching was infrequent, but that is sometimes the wages of true adventure. And as long as I live I will never forget that single lovely fish I released that day, or how glorious the coffee tasted, or how brightly we laughed together as we joked about the weather. If every day were easy, what would be the point? It is in the struggle that we are all defined.

We all come into each day as we are, and as we choose to be. I have learned over the decades of life and living to simply accept each moment as it comes to me with as much grace and gratitude as I can. Fishing is my teacher. I guess I need to become a bit more dedicated at perfecting more varieties of "perfect casts and presentations." I'm sure I need to learn how to tie more types of knots and get better at tying my own leaders and so many more skills that I may lack as an angler—and I will.

Part of what I love about fly fishing is that I'm always learning, and in doing so, I'm always living. But what Nature and fly fishing really teach me is how to live a balanced and mindful life. I learn about patience, persistence, and perspective. I learn about acceptance, impermanence, and interconnectedness. I practice letting go of expectations and limitations that exist only in my imagination while embracing the magic of each moment—however it may unfold. I can do nothing about the rising trout, only my response to its choices. Most of all, fly fishing and being fully present in Nature reminds me to never miss a chance to experience joy. And there is always that chance.

I still wish my friend had connected with that final rising fish, but I also hope he enjoyed every cast, drift, and rejection. When he dropped me off at the inn, it was quite late and as dark as night can be. I walked into the doorway by the kitchen, making sure that I was as clean and dry as possible out of respect for my new friends, the innkeepers. Almost as soon as I came in the door, Joe walked into the kitchen with a big smile on his face and poured me a glass of wine. He handed it to me and asked, "How was your day on the river?" I smiled back, raised my glass, sipped the deep, organic ripeness of the red wine, and said, "My day was perfect, my friend . . . absolutely perfect." And you know what? It was.

Chapter Sixteen

Wading the Esopus, New York

Home after all, is more than a place on a map. It's a place in the heart.
—J. Drew Lanham

When I first stepped into Esopus Creek, I had no idea where I was standing or what it would come to mean to me. Todd Spire is the program and events coordinator at the Catskill Fly Fishing Center and Museum. He is also an avid angler, guide, historian, conservationist, and my friend. My buddy Joshua Caldwell introduced us to each other, and it didn't take long for the three of us to plan a trip to explore Todd's home water—Esopus Creek.

Josh and I followed Todd's lead as we drove up a winding woodland road and pulled into a dirt turnout somewhere at elevation above the forested valley. When we stepped out of our vehicles all I saw were trees, but Todd assured us that there was running water with native brook trout just inside the tree line and over what turned out to be a relatively steep embankment. It was steep enough that someone had previously had the good sense to tie a long, thick rope to a tree along two sections so that you could hold on to the rope with one hand and your fly rod with the other as you carefully worked your way down toward the small waterfall and wide plunge pool that waited below. We could see the little brookies in the pool even before we made it down there—and by little I mean that a ten-inch fish was a lunker.

I was graciously given first shot at the pool full of little native brook trout in this postage stamp–size paradise where the memories of the paved road above us had already faded. I've always been surprised about the resilience and diversity of the eastern forests. It so often seems that with just a few steps down a trail, across a meadow, or over an embankment you can find yourself in a magical world of maple trees, babbling brooks, and salamanders that always seem to be smiling. But these verdant forests of maple, oak, and birch trees also offer the added challenge of casting between the same branches that filter the sunlight and shelter the warblers. So I lined up my cast and methodically began working the little hopper-dropper rig into the current and over the brilliant little fish. They remained unimpressed.

Josh had already moved on and was fishing somewhere downstream and around the bend, but Todd and I lingered at the first pool where I picked up one missed strike and dropped the one opportunity these wild trout were allowing me. We worked our way slowly from pool to pool without result—at first. Then Todd and I came to a fork in the stream, and we could see that Josh had taken the left fork and was in the process of releasing a pretty little brookie. And one by one we each began to unlock the mysteries of these wild, native fish—catching and releasing them back to the tiny plunge pools that for them contained the entire universe.

We're no different, really. The world as we know it exists only in the immediate vicinity of wherever we're standing or inside the magic screen of our iPhone. Sometimes that magic is more sorcery. It depends on the click of a cursor or the algorithms that carry information and misinformation to you, like so many currents in a mountain stream.

It was peaceful under the forest canopy and along what appeared to be an unnamed Catskills creek. There was no indication that the anxieties of "civilization" might be found at any moment, by happenstance or by choice. As relatively small as this woodland water was, it also contained the power of gravity and the powerful ability to drown not only grasshoppers and crickets but also any sounds other than that of falling water and resisting rocks. At one point along the stream, I climbed a steep hill that overlooked a wide pool below and a series of waterfalls and pools above.

I looked down to see Todd fishing the pool and over to see Josh fishing the pocketwater between the falls. Both seemed to be intensely focused on the mission at hand—namely, fool a small, wild trout into thinking a bit of fluff on a hook was a meal, catch the small fish, and then release the small fish.

Sometimes the absurdity of my favorite outdoor activity makes me smile. And I was smiling at that moment as my friends caught and released fish, but I simply stood there looking and listening to everything around me. I wondered if they noticed how beautiful the falls were, or how the leaves were waving in the breeze. I wondered if they noticed the call of the blue jay or the warble of the warbler. And I wondered how much of this I missed when I too placed all my attention into the few feet of water that suspended a wild fish and an older man's younger dreams.

We humans are inclined to think of history as linear even though Nature and the nature of the universe teaches us that everything is a circle and a cycle—never ending or beginning, and always in existence in one form or another. Water flows from the sky to the rivers and from the rivers to the sea and then back into the sky—and along the way it contains and sustains the rising trout that sustains us. Water is life, and life flows freely without end.

Esopus Creek and its tributaries seem to almost unnaturally form a nearly perfect circle. Unlike most of the earth's rivers, the Esopus does not seem inclined to flow toward the sea. It lives in circles and cycles of both form and function. By way of form, rivers can only truly be understood with a microscope or telescope focused deep inside each drop of stream water or far above the earth's landscape. Our existence is like that also. To see its form more clearly, we must step away. And when seeking to understand its function, we must take a more spiritual and intimate approach to knowing a river—we must step inside it and become one with it. Only then can we come to know the river's true nature.

New York State geologist Yngvar Isachsen was taking advantage of the faraway view of the Esopus as he tried to discover why the valley of the Esopus and its tributaries formed an almost perfect circle around Panther Mountain. He was working on a NASA-funded project that sought to better understand the geology of the Catskill Mountains, and as he pondered the circular nature of the Esopus, he asked himself, "What on earth would account for that?" He concluded that what caused it wasn't from the earth. It was the result of an impact crater from a meteorite that struck some 375 million years ago.

While a comet is made up mostly of ice, a meteorite is rich in iron. Droplets of that iron were discovered by Isachsen's team all around Panther Mountain and the Esopus basin, as were the shock lamellae of melted quartz caused by the impact of the meteorite. But this one amazing fact is just one piece of the many that make this river and these ancient mountains so special. A lot of history happened here, and we are living in a time when much of the future of this historical landscape will be decided.

We continued to wander with the stream and within the woodlands that surround it. I had no idea where we were and was putting all my trust in Todd. We had each caught a few fish and more than anything I was feeling content and happy about the place and manner in which Todd had chosen to begin our morning. That's when I asked him, "Does this little creek have a name?" He smiled and said, "This is the headwaters of the Esopus." I was thrilled.

Catskill Park consists of 700,000 acres of forest preserve that acts as habitat for a wealth of wildlife and is the main source of drinking water for New York City and its 8.5 million people. Within the Catskills resides the 47,500-acre Slide Mountain Wilderness, which is the largest state-owned and legally protected wilderness area between the Adirondacks and the southern Appalachian Mountains. I am grateful for publicly protected lands and desirous of many more places to be legally taken out of play for developers and corporate exploitation. My own home state of Texas has an abysmal lack of public land, and I for one would like to see that change in my lifetime.

These forests that have managed to survive in the shadow of New York City contain almost fifty species of mammals including black bear, cottontail rabbit, snowshoe hare, gray squirrel, porcupine, and beaver. White-tailed deer were once extirpated by the advance of early

Euro-Americans but were successfully reintroduced into New York in 1887. The fisher was also shot and trapped to extinction in these woods along with the once native eastern elk and mountain lion. The fisher was reintroduced to the Catskills in the 1970s and is slowly making a comeback. The eastern elk and mountain lion had no biological reserves to draw from for reintroduction, so sadly they are gone forever. While states like Pennsylvania and Kentucky have successfully reintroduced elk, they are a Rocky Mountain subspecies and, to my mind, a poor substitute for the original. Painted horses can't replace zebras.

Catskill songbirds include the threatened Bicknell's thrush, as well as the black-and-white warbler, black-throated warbler, Canada warbler, red-eyed vireo, scarlet tanager, rose-breasted grosbeak, and many more. This is also home to native game birds including ruffed grouse and eastern wild turkey. Almost one hundred species of birds rely on these forests for habitat during at least part of their life cycle. And of the thirty-seven species of amphibians and reptiles found in the area, six have been designated as requiring special concern and protection including the Jefferson salamander, blue-spotted salamander, spotted salamander, spotted turtle, wood turtle, and eastern hognose snake. As an angler and naturalist, I am aware that the fish I seek to find are forever connected to the entire ecological community—and so am I.

The forest that surrounds Panther Mountain and the Esopus watershed is a mix of beech, yellow birch, sugar and red maple, white ash, red oak, basswood, and big-toothed aspen along its slopes, with remnants of white pine and eastern hemlock near waterways. These northeastern hardwoods give way to boreal forests of balsam fir and red spruce above 3,500 feet. At 3,720 feet in elevation, that leaves a sliver of remaining habitat for these cold-loving trees and the various birds and other wildlife that depend on these high coniferous islands in a sea of hardwoods. As global temperatures continue to warm at an alarmingly unnatural rate, creatures ranging from crossbill finches to cutthroat trout will continue to vanish as too many of us look away in denial and apathy.

It has been written that in the final moments of life, Jesus of Nazareth spoke the words, "Father, forgive them . . . they know not what they do." I'm not sure if he said this or not; I wasn't there, and if you are reading this, neither were you. But I am sure that I'm no saint and I think he was letting us off easy. The truth is, we do know what we did and what we're still doing. Some of us are driving the nails forward while some of us simply look away and mumble, "That's a shame." Well, it is. As I stood next to the tumbling headwaters of this ancient stream, all I could think was, "I don't want to live in a world without wild brook trout and free-flying songbirds filling the waters with poetry and the treetops with music."

So much human history has unfolded in these ancient hills. After our short foray along the headwaters and the chance to catch and release native brook trout, we moved down the watershed for a brief stop at Schoharie Creek, made famous by lifelong angler and stream conservationist Art Flick. After a careful crossing of the Schoharie's fast waters, we each took our positions along what was reportedly one of Art's favorite sections. Todd and I tossed traditional nymph setups and Josh broke out his Euro-nymphing rig. I watched as he worked the same waters I was, except his efforts ended in the landing and releasing of a nice brown trout, and mine ended in a bit of casting practice and a pleasant moment on a historical stretch of northeastern water. While it's true that every landscape is a historical landscape and that no place on the earth has remained untouched by human influence, these waters and this landscape are uniquely historical.

In his book titled *Catskill Rivers: Birthplace of American Fly Fishing*, Austin M. Francis writes that the Esopus was given its current name by the Dutch settlers who took it from their version of the Algonquin word for "small brook." The Algonquin-speaking tribes were the first known people of this region. Prior to being pushed out by Euro-American settlers, they lived in and hunted the forests surrounding the Esopus, Beaverkill, Willowemoc, and Neversink Rivers. Whereas the Euro-American immigrants brought with them Calvinist ideas of subduing Nature, the Algonquin culture lived as a part of the community of Nature. There's a lot we can learn from the First Nation peoples of this beautiful continent.

The Catskill Mountains region is often referred to as "The Birthplace of American Fly Fishing," and for good reason. The list of iconic anglers, innovators, and conservationists here is long and legendary and includes names such as Paul O'Neill, Theodore Gordon, Arnold Gingrich, A. E. Hendrickson, custom rod maker Jim Payne, Joan and Lee Wulff, and innovative fly tiers Tom and Chester Mills of the New York City tackle store William Mills & Son.

Perhaps one of the most fascinating life stories for me is that of the great Joan Wulff, who was a pioneer in bringing more women to the fly-fishing lifestyle. As a young woman, Joan learned to cast via lessons from her father and with the help of instructors at the Paterson Rod & Gun Club in her hometown of Paterson, New Jersey. In 1978 she and her husband Lee Wulff opened the Wulff School of Fly Fishing along the Beaverkill River. The positive impact that one person can have on the lives of others is immeasurable.

By lunchtime we had moved down from the headwaters to a place along Esopus Creek where an old railroad track ran, unused, unnoticed, and largely covered in leaves and acorns. It felt poignant, and I was compelled to take some time walking along it as my friends enjoyed sitting by the river. I didn't go far since there was nowhere to go. At one time this track carried tourists and commerce from New York City to the cooler mountains of the Catskills in summer. And before that time it was built by the labor of bodies long since defunct and decayed in some graveyard that nobody visits, because nobody is alive who remembers them. I don't know where the essence of their blue-collar souls ended up, perhaps in oblivion, perhaps in the sapwood of the surrounding trees. It matters not.

Our lives come and go, and once they're gone what remains is what we've left behind in words and deeds. We can leave behind railroad tracks or forest. We can leave behind greed or generosity. We can be known for hurting or healing. It's a choice.

Things were slow at the lunch place and, as pretty as the river was, none of us got so much as a nibble on anything we drifted downstream. So we packed up our gear and left the ghost-train tracks behind while we followed Todd once again farther downcurrent and along the lifeline of the Esopus. Where we began, the Esopus was small enough to step over and full of wild and native brook trout that were naively eager to take a fly. At the lunch spot, the impact of humanity included railroads and roadways and the absence of native fish. This was rainbow trout country, and the creek had become a stream, perhaps thirty to forty feet wide. Now we worked our way to the place where Esopus Creek seemed more like a raging river, thirty to forty yards wide. Here the river could drown you if you let it, and the sounds of the roaring water certainly drowned out the sounds of "human progress."

When I asked Todd why he loved the Esopus and felt it was his one true home water, he spoke of having a relationship with the river, mountains, and forest and how he grew to miss these things whenever he was forced to be away from them for any amount of time. After waxing poetically about the river and its surrounding landscape, he spoke practically of what he found to be so special about Esopus Creek. He said, "One of the incredible aspects of the Esopus is that one need not fish it reactively. By that I mean you can often wake up and decide how you want to fish and simply seek out the spot to fulfill your desires. From its tiny headwaters to fast pools and pockets, the river offers nearly every option for anglers, from streamers to dries and everything in between."

The one thing that I have discovered about so many eastern rivers is that they are just over the guardrail of way too many busy roads. But here, we had to shout to be heard over the roar of the water, and the sounds of traffic were a distant memory. Here, we were lulled into a peaceful state of awareness, and the only thing that existed in the universe was the loop of our line, the drift of our leader, and the exhilaration of leaping rainbow trout—of which there were plenty. We were all catching fish.

The water was fast and furious in this stretch of river, with large boulders interspersed between slick, round rocks that ranged from golf to soccer balls in size. So for the most part, we either cast from the bank or waded ever so carefully into the swift currents. Todd was downstream of me, and Josh was upstream, and with a good bit of regularity we were connecting with eager rainbows that ranged from twelve to sixteen inches in length and, more importantly, were bright, torsional, healthy, and completely wild. Like me, these fish had no idea that they didn't belong here.

If I'm being truthful with myself—and I try to be—I wish that only native fish prevailed in these waters, not the fish that were trucked here from seaports long ago. I wish the Adirondacks, Catskills, and

Appalachian Mountains were still brook trout country, and the Rocky Mountain West was still cutthroat country, and the Pacific Northwest and Alaska was still the only place to go for rainbows, steelhead, and Pacific salmon. But that's not the reality we've created.

I suspect that quite like we've fashioned any deity in our own petty image, we've inadvertently changed North America into the multinational reflection of its human population. Just as there are people of every kind of previous origin, race, and religion who call themselves Americans, we have altered our ecosystems irrevocably—pushing out the native for the non-native and the natural for the unnatural. Trees became trestles and the Algonquin were supplanted with visiting members of the Algonquin Round Table.

I am torn by my realization that while the natural world that once was here must have been indescribably beautiful, what remains of these forests and streams has its own miraculous charms. After all, neither the rainbow trout nor I bear any ill will toward these woodlands. We are kindred spirits—misfits transplanted by the choices of a previous other. We are blameless—yet significant in our insignificance and impermanence.

Whenever I catch and land a fish, I make a point of looking into its eyes and recognizing its aliveness. One eye looks at me as I hold its life in my wet hands. The other eye looks toward the water that allows it to live, just a little longer. And as I watch it swim away, I feel relieved for the chance to undo what I've done. Perhaps one day I will simply walk along rivers and look for the trout, never casting my line. But on this day standing next to my friends, I cheered every time they hooked and landed a fish. And I radiated joy and gratitude for every leaping trout I managed to bring to hand. We're a strange species—humans.

The Catskills are a geologically and culturally historic landscape that contains the memories of glaciers and grand migrations of humans, first from the far eastern lands of Siberia across the North American West and then from the far western lands of Europe across these verdant forests toward the Great Plains, Rockies, and Pacific Ocean. Change is our only constant, and it's perhaps ironic that the same people lamenting the current massive influx of immigrants are like me and you and the Potawatomi—immigrants, one and all. Go back far enough in time, and we are all travelers from the African savannas, just a few steps from the cave and a geologic blink from hiding in treetops.

This is not to say that the migration of *Homo sapiens* from east or west or the forced migration of trout and trees is without a price. When the first humans crossed the land bridge into modern-day Alaska, they brought with them tools and technologies that led to the mass extinction of most of North America's megafauna. When the first humans crossed the Atlantic Ocean, they brought with them tools and technologies that led to the mass extraction of fossil fuels that are forever changing the climate and chemistry of our planet—atmosphere, soil, and water. We brought swine and cattle that now fill our arteries with cholesterol, our rivers with organic waste, and our atmosphere with methane. We tossed European brown trout in our rivers and Mesopotamian grains across our prairies. And now the planet is burdened with over eight billion humans, each and every one of us wanting "more."

Yes, there is a price we extract for our continued presence on this beautiful blue marble. But it doesn't have to be this way. We can turn the wheel and move in new directions. Brookies and brownies, Bostonians and Brahmans can each have room to live, without extinguishing each other.

In her landmark book *Rambunctious Garden: Saving Nature in a Post-Wild World*, author Emma Marris makes the case for a hybrid of wild Nature- and human science–based management of the natural world. She argues that the damage has already been done, and mitigation and recovery are the new plausible goals. In his book, *Nature's Best Hope: A New Approach to Conservation that Starts in Your Yard*, author Douglas W. Tallamy argues for the transformation of our living spaces—urban, suburban, and rural—from heat-holding concrete, worthless lawns, and monoculture fields to interconnected "nature-scapes" of native trees, shrubs, flowers, and grasses. I see this as a good direction for every angler, hunter, hiker, paddler, and other outdoor enthusiast who loves the magic of being outdoors to get behind. How about you?

I'm not deluding myself. We won't be replacing every brown trout in Montana with a cutthroat or every rainbow trout in New York with a brookie. All I'm saying is that our native fish and other wildlife need to have places to survive and thrive. Recently, my friend the conservation journalist Ted Williams was asked, "Where do we draw the line about where we regain and sustain native fishes and where we allow the non-natives to live—wild and free?" I thought his answer was spot on target. He said, "I think Nature gives us that answer." Just as there are big waters where we will never be able to eradicate non-native fish, there are places where we can and should repair the damage we've caused and create the conditions for native species to live and flourish. To my mind, native plants and animals are both our teachers and preachers. They teach us the natural way of things, and preach to us the gospel of empathy, understanding, respect, and responsibility. And to those lessons and homilies I say, "Amen."

It was getting late, and the light became increasingly dim as we cast and caught while standing side by side on this beautiful and historic river. My eyes are not as clear as they used to be, and the waning light made watching indicators and tying on flies challenging indeed. But no amount of evening darkness obstructed the visions I was seeing in my imagination. I could clearly see my indigenous brother standing across the river. It didn't matter if he was Delaware, Mohican, or Mohawk, he was clothed in hides, fur, and feathers. By covering himself in the artifacts of Nature, he signified how deeply he was connected to it. As I made one last cast, I considered the fly at the end of my line. It too was clothed in hide, fur, and feathers—suggesting that it might be a bit of trout food and, in doing so, bring Nature closer to me.

It's not that I wish I was that last Mohican of long ago. I must admit that I enjoyed the feeling of the heater in Josh's SUV as he drove me back to the Inn at Stony Creek. I was happy to stand in the kitchen chatting with Joe the innkeeper, sipping red wine and sharing news of our day. And the warm shower and comfortable bed probably felt better than any native wickiup or cabin, but still I get the feeling there is something they had that we've lost.

When I first stepped into the waters of the Esopus, I felt I was beginning a wonderful journey somewhere new, but I also knew that I was just passing through. And therein resides the difference between my indigenous friend and me. He was "Home." Being Home makes all the difference in the world. Like the rainbow trout, I am an interloper, a dreamer in a foreign land, by no fault of my own. And I get the feeling that like the rainbow trout, I could feel at home here in these wooded mountains. I could learn the intimacies of this place and make it my own. And as I grew in my love for these waters, I would grow in my caring for them and their future.

So I'm still wondering, "What is Home?" I don't yet know. Perhaps I will, if I keep casting forward with narrow loops and an open heart. I hope so. Now more than ever, I need to be Home—whatever that is and wherever it may be and however this world might change. Perspective is so vital in feeling either like a refugee, traveler, or new native in any place and time. If only I could afford that Airstream trailer and new pickup truck to pull it, I might always be home . . . or not.

Chapter Seventeen

Wading the Farmington, Connecticut

"The presence of those seeking the truth is infinitely to be preferred to the presence of those who think they've found it."

—Terry Pratchett

Throughout this entire journey across North America, from Alaska to New York, I have begun with the question: "What is your home water and why is it the home of your heart?" I wanted to boil down the sauce into a roux of pure understanding of what makes someplace feel like home and, in doing so, an extension of our spiritual being. For my spiritual brother Bob White, the Wood River drainage was where he had guided and fished for formative decades. It was there that he made lifelong friendships and met his amazing wife, Lisa. The rivers within easy driving distance of Bozeman are the ones connected to the beautiful life my friends Sue and Josh have created together. And the streams and ponds of eastern Oklahoma are where my friends Emily and Dave Whitlock loved and lived together until Dave's passing on the day before Thanksgiving 2023. I think I see a pattern of understanding unfolding here. Home really is where your emotional and spiritual heart lives.

The idea of our spiritual being is interesting in itself. My buddy Ross Purnell and I are both big *Star Trek* fans. We've shared our observations of how ahead of its time and prophetic each show has been and what we have learned from them over the years about humanity in all its potential forms. Just the other day, while watching an episode of the original show,

I got to wondering, if the transporter rearranges your molecular structure and then reassembles it somewhere else, how does your soul travel with you? I've thought the same thing during my past life of being a Marine counterterrorism specialist and Texas Master Peace Officer who has seen a lot of violent death up close and personal. I've wondered, when the light leaves our eyes, where does it go? Where do we go?

I no longer believe in any celestial heaven or subterranean hell. I believe that we create those fictions and realities in our minds and through our choices. Still, I also doubt the idea of our essence drifting into oblivion since the nature of Nature is renewal. In life and in fishing, we never know what comes next around each bend, but in time it is revealed to us. Eventually, we all drift around that bend in the river of life, and in doing so, we are forever changed.

A few months before taking this trip, I wrote to my friend Joshua Caldwell and asked what his home water was. He didn't hesitate and said, "the Farmington, in western Connecticut." I asked why, to which he replied, "There are rivers you fish, and then there are rivers you come back to, time and again, like a favorite old book you can't help but reread. The Farmington falls into the latter category for me. It pulls at me. And there's something brutally honest about the Farmington. Every cast is a gamble, filled with possibility, and if I walk away empty-handed, it's on me. The river's always teeming; it's just a matter of whether I've got what it takes on that particular day."

Josh and I had fished together before on the Esopus and Delaware Rivers near his beautiful home in the Catskills of New York. But these were the rivers closest to Todd and Landon's hearts. This was my first chance to fish with Josh on his favorite river, a river that he had to invest the time of a two-and-a-half-hour drive into Connecticut to fish, even though other fine rivers flowed much closer to "home." Whenever I travel to the waters where a friend might feel most intimately connected, I don't necessarily come away with a new home water of my own. But I often do gain a better understanding of my friend, and perhaps of myself. You might think that a man in his sixties would have accomplished enough discovery and learning to fill a lifetime, but I'm just getting started. I never want to be "set in my ways" or an "old dog." I'm always searching to learn new tricks.

In the few days we'd been fishing together, Josh and I had created our own comfortable traditions. One that might not have been so good for my wounded heart was certainly healing to my wounded soul. Each morning Josh picked me up at the Inn at Stony Creek, and we'd stop at the deli to pick up our lunchtime sandwiches and breakfast sandwiches. I had decided to live dangerously and allow myself to enjoy every decadent egg and cheese moment, even though my efforts at home to maintain a heart-healthy diet might suffer a little. And since this day was especially special, we bought three breakfast sandwiches, so that we could each have one and a half moments of breakfast nirvana with our glorious morning coffee. It felt wonderfully conspiratorial as we drove through the verdant green countryside of the Catskills toward the waiting river while devouring every fat- and cholesterol-filled morsel. I have sinned, and I'm not contrite in the least about it. In fact, I'm likely to do it again.

Growing up in the American South, I always saw the Northeast as a faraway place with somewhat foreign inhabitants. I mean, they say things like "you guys" and seem to always be in a hurry to get somewhere. I've since learned that these fine people from these wonderful places found us Southerners just as perplexing. Back then I was accepting the stereotypes that come with a lack of actual intimate knowing—and so were they. In this part of my life's journey, I have found some of the most amazing human beings and lifetime friends in places like New York, Pennsylvania, Massachusetts, Vermont, and Connecticut. And I have developed a deep affection for the forests, rivers, shorelines, and communities of the northeastern states of this all too often "un-United" States. I have found what I should have always known—that in both our light and darkness we are more the same than we are different.

Driving through the Catskills of New York or the Green Mountains of Vermont or the forested landscape of western Connecticut, I find myself feeling oddly at home. This southern boy has come to love the American Northeast and its people. I care about these places and the many faces who have smiled at me, accepted me and the way I say "y'all," and treated me kindly. We need more of that in this world. Don't you agree?

We arrived at the Church Pool of the Farmington and met up with Josh's friend Steve Hogan, who among other things is an excellent fishing guide who specializes in the rivers of western Connecticut. When we first pulled into the parking space, I jumped to a conclusion, which is something I know better than to do. You see, I immediately noticed that Steve was speaking to Josh but I seemed left out of the conversation, so I figured it was going to be a long day of being invisible. I was so wrong. It was just part of the slower pace of getting to know each other that comes with the cultural territory. As it turned out, Steve Hogan would quickly earn a special place in my life's journey as a good man with a natural ability to teach and a wonderful ethic for angling. I'm once again so grateful to Josh for introducing me to truly wonderful people. Everywhere I've traveled around this planet, I've met amazing people of kindness and humility. A mutual love of being immersed in Nature is often the common thread that connects us.

Once again I noticed that both Josh and Steve were geared up and ready to go with such efficiency and velocity that it seemed they must have dashed into a phone booth and come out fully clothed, like angling supermen. I was definitely seeing a cultural trend. Perhaps this rush to the river comes from the fact that here in the Northeast, if you don't get there quickly, someone else might be there already. When I fish back in Texas, unless I purposely go to a place I know is "popular," I will not see another angler on the river—ever. There's no rush. Also, bass don't care about hatches. They will eat anything that moves that might fit into their gaping maws. In fact, it can be a good thing to take your time and allow the water to warm a little while you sip your coffee and sit on the tailgate listening to the birds singing and watching the sun rise. I understand the origins of stereotypes, but the reality is that down south we're not lazy, we're just lucky. We can take our time and sit on the tailgate catching up while lying back, and that's not so bad, y'all.

In past fishing adventures with Josh, I've watched him take part quite successfully in the art of Euro nymphing. He seems to always carry two rods with him wherever we go—one for dry flies and streamers and another for Euro nymphing. Anglers in places like the Czech Republic, Poland, and France perfected the technologies and techniques that are now commonly referred to as Euro nymphing. Rods that have been developed for Euro nymphing tend to be longer than normal fly rods, ranging from ten to eleven feet in length, thus allowing the angler to reach farther with the short, upstream lob presentation that is a hallmark of this style of fishing. These extra-long rods tend to have thick butt sections and fine, ultra-responsive mid and tip sections that allow the angler to feel every tick of the bottom and take of the fly. The line setup includes six to twelve feet of clear mono, a sixteen-inch section of brightly colored sighter line as an indicator, and several feet of level fluorocarbon tippet to the fly. Instead of casting the weighted line up and out for distances of twenty to sixty feet or more, the Euro-nymph angler uses the weight of the nymph to reach the target water, which is often within ten to fifteen feet in front of the angler. You can only cast as far as the rod and leader reach, so this is an intimate way of fishing. I had read a few articles in *Fly Fisherman* magazine about Euro nymphing and watched Josh using the technique on the Esopus to great effect, so today I wanted to give it a try.

Josh tried to teach me the process as we stood on the bank of the Church Pool, but my color blindness did not allow me to see his pink sighter line, so Steve offered me his rig and what I can only describe as patient and perfect instruction in both the mechanics and the nuance of Euro nymphing. One of the aspects of this technique that I immediately and naturally gravitated to was that it allows me to methodically, thoughtfully, and intimately explore each section of likely habitat on the river. The first order of business is to watch the river, read the water so as to intuit what the riverscape might look like and where trout are most likely to hold, and then select a particular run to be your first focus of attention. Then, after lobbing a cast upstream, you raise the rod tip and tighten the line while conducting a timely follow-through, moving the rod tip with the speed of the current. Almost immediately I felt myself getting into the groove of it all, and after just a few passes I connected with a pretty little rainbow trout—my first fish on the Farmington. I've heard other anglers disparaging this style of nymphing as "not real fly fishing," but for me, I had experienced an angling epiphany! I felt intimately connected to the river and the fish. It was primal.

Steve saw how much I was enjoying it, and when I tried to give him the rod back he said, "No, no, just go ahead and practice. I can fish here anytime I want." So I did, and before long I picked up and released another smallish rainbow trout and missed a couple probable takes. I was still feeling a bit awkward as I adapted to this style of fishing, but in a way, it felt quite natural. I had flashbacks of my cane pole–wielding childhood and began to wonder if I need to finally break down and try a Tenkara. For years now I've contemplated getting one of these simple, extra-long, telescoping rods that uses a fixed-length line to deliver a single wet fly to the water, but I have not done so yet. Perhaps this will be the year.

Although we each caught a few fish and I enjoyed experiencing one of the legendary pools along this beloved river, things were a bit slow at the Church Pool. The water was cold and fast and had already been fished by a couple of anglers who were in the area when we arrived. So, after a short conference between Josh and Steve, we picked up our gear and drove to the next section of the river—and I'm so grateful that we did. I decided to call this stretch the "Bear Woods Pool."

To get to the Bear Woods Pool we had to drive down a dirt road, park at its terminus, and then walk a mile or so down a woodland path that was shaded by hardwoods and serenaded by warblers. As you might expect, there is a reason I decided to call this the Bear Woods Pool. As we walked off the trail and toward the river, Steve told us the story of a recent time when he was fishing here and enjoying the solitude and peace that comes with having a section of forest and river all to yourself. He had been quite engrossed in catching and releasing a few fish between intermittent puffs of a fine cigar when he felt a presence, as if someone was looking over his shoulder. It was a bear. Not an angry bear, just a curious bear. I loved that story, and the fact that bears have rebounded in western Connecticut makes me love these forests and this river all the more.

While the Church Pool might be legendary, the much smaller and wilder runs of the magical place Steve took us was by far my preference. I love smaller and more intimate water, and this was also the perfect place for me to continue my Euro-nymphing experience. Steve's kindness, generosity, and patience were astounding, and made all the difference. He and Josh altruistically gave me the run that they knew was most promising, and as it turned out, those promises were kept. Josh began fishing just downstream from me, and it didn't take long for him to connect with a nice fish. At first, Steve decided to stand on the bank, enjoying the scenery and a cigar while he gave me expert advice that I took to the best of my ability. It just felt right.

I began my first pensive flip-casts upstream and practiced following the current, feeling the tick, tick, tick of the fly crossing the river bottom and then the soft pause of a fish taking the fly. With just a twitch of my wrist, I was connected to a beautiful fish that fought briefly and then came to the net. This was a brown trout and a much bigger fish than the little rainbows I had encountered in the Church Pool. I held him briefly at the water's surface as Steve took a quick photo so I'd remember every detail of the moment. After that, I would have concluded that the run had been spooked, but Steve told me to just take a step up- or downstream and try again. By the time I had completed working the entire run, I had landed half a dozen fish. Euro nymphing is effective to say the least.

After fishing that entire seam of current, I simply stepped into the water I had been fishing and began working the next run on the far side of the narrow side branch of the Farmington. Josh had picked up a few more fish, and Steve picked up Josh's other rod and fished upstream of me as I monopolized his expensive and unbelievably sensitive Euro rig. I was having so much fun, but not just because I was catching fish—it was more than that. This felt somewhat Zen in nature, which is right up my alley as an Imperfect Texan Buddha and Warrior-Poet. There was a sense of oneness to this endeavor and a peaceful calm that might have passed for meditation.

I began experimenting with closing my eyes and seeing if I could "see" the river by just feeling it at the end of my tight line. I'd try to move the rod tip with the current using my sense of feel only, and then differentiate between the tick-tick feeling of the fly hitting the river bottom and the living-being feeling of a fish mouthing the fly. And it worked. Before we wrapped things up on this little stretch of water, I caught and released a few more fish—and I was catching them intuitively, not mechanically. This is what I loved about it, the deeply personal connection. I felt the river and the fish. I experienced the river's current not by passively watching a strike indicator float past me but by actively staying connected to everything. And much to my semi-animistic delight, the river felt alive, and the fish wasn't an inert object being yanked from the water. It was a living, "breathing" being, and we were, ever so briefly, connected to each other. Yes, I realize that this connection wasn't as pleasant for the trout, but I also realize that if these fish, waters, and forest are ever going to survive humanity, we humans need to connect with them—intimately.

I felt almost melancholy as we walked back down the path through the dappled sunlit woodlands and toward the clearing where we had parked. The day of fishing wasn't over, but I couldn't help but wonder if I'd ever see the Bear Woods Pool again—except in my dreams. It's not that the pool or the river suddenly became my new home water. It's that I felt at home in this water and walking through these lovely eastern hardwood forests. I also enjoyed the company of my friends Josh and Steve. I even loved the aroma of their cigars and wished my asthmatic lungs and fear of lung cancer would have allowed me to partake with them. Like a little boy at a bluegill pond, I just didn't want the day to end. All I could feel was utter ecstatic joy.

The west branch and main stem of the Farmington River is heavily stocked with non-native brown, rainbow, and native brook trout by the state of Connecticut, but is reported to have about 40 percent of its trout population as wild and self-sustaining fish. The west branch between Barkhamsted and New Hartford is a bottom-release tailwater where the river remains cold and oxygenated even through summer, although I have recently seen ever-increasing warnings online that water temperatures are rising dangerously in summer as climate change causes more and more days of excessive heat each year.

As anglers, we need to stay vigilant and refrain from fishing and causing increased fish mortality on the hottest days. As a naturalist, I'm hoping we realize as a community that we must protect the wildlife and habitats we profess to love. There may come a time in the near future when many of our so-called legendary trout waters can no longer sustain wild trout. I'd like to avoid that time.

The west branch of the Farmington River is designated as a Wild and Scenic River by the National Park Service. As an angler and a naturalist, I realize that a river is not a "fishery" any more than a forest is a "lumbery" or a prairie is a "beefery." Wildlife and wildlife habitats are not our resources to exploit; they are our neighbors to be respected. We need each other.

Rivers and forests are interdependent on each other and sustain life as a community. On either side of the Farmington are public lands in the form of state forests that provide habitat for black bears, white-tailed deer, moose, and wild trout. Without a healthy riparian habitat, rivers become nothing more than empty waterways. Public lands often make all the difference in the preservation of healthy rivers and streams. In this case, the 3,059-acre Peoples State Forest protects the east bank of the west branch, while the American Legion State Forest provides shade, stability, and nutrients for the west side of the river. And, of course, even though we humans feel the need to create imaginary dividing lines, this entire woodland is part of the northern hardwood and oak-hickory hardwood types of North American forest. The trees that surrounded and shaded us were red and sugar maple, black and yellow birch, northern red oak, eastern hemlock and white pine, beech, black cherry, and much to the delight of squirrels, a few pignut hickories.

Every forest contains multitudes of plants and animals, microbes, and fungi, each connected to the others in intricate and important ways. And each river is just as complex in its aquatic community. I never forget this as I walk into a forest on my way to a river in search of wild and hopefully native fish. It's why I prefer to go slow, flip a stone, and see who's living underneath, and then gently replace the stone that is someone's home. For me, fishing will always be more than just catching fish, just like hiking is about more than getting from one place to another in the least amount of time possible.

The Ovation Pool gets its name from its location near an old Ovation guitar factory. Once we parked and got out our gear, we had a short walk through a bit of boggy woods before we came to a wide-open, flat stretch of the river with several nice braids of fast, fishy-looking water. At the point where the trail ended and the river began, there was a small wooden bench that seemed inviting, and I wondered how many anglers had sat there over the years looking across the pool in deep thought or, perhaps, without thinking at all. Sometimes these places cause us to search for our soul; sometimes these places cause us to find it.

In the American West I am often grateful for the lack of human constructs along the rivers, which gives the impression of open space and solitude. In these wild places, we find connection with Nature. In the American East I am often comforted by the streamside benches and riverfront gazebos that remind us that we are not alone, and in that knowledge, we find connection with each other and the best of human nature. We need both worlds.

At first, Steve and I waded out to a curved bit of fast water, and he gave me a few more valuable lessons about this particular style of angling called Euro nymphing. It was a bit more challenging here, but in time I began connecting with a few trout. Josh was doing the same downstream of me. As much as I was enjoying the use of Steve's gear and expertise, I insisted that he take his rod back and have a little fun himself, and I began casting my own nymph under a bobber rig. Although I did catch a couple more fish, it just didn't feel the same.

It was fine. I'd had enough and Steve managed to catch a small Atlantic salmon on the Euro-nymph rig, so we all decided this was the perfect moment to snip off our flies and call it a magnificent day. I never count how many fish I catch or how big they are because I really don't care, but we caught enough rainbows, browns, and one salmon to put this trip "in the books" as one I will never forget. Still, the thing that stood out the most for me wasn't the fishing or even how lovely the forest was, but rather the friendship we shared along that wild river and thriving woodland.

On the way back to New York, the Catskills, and the inn, Josh and I chatted about the day we'd shared and how much I enjoyed my first try at Euro nymphing and my first visit to his favorite river. We made a quick stop at the Croton River, a favorite haunt of the iconic outdoor writer and originator of the company that publishes my books, Nick Lyons. I have admired Mr. Lyons's writing for decades, and amazingly I now have the pleasure and honor of being the dedicated author of the "Seasonable Angler," his former column in *Fly Fisherman* magazine. Life unfolds in so many mysterious and magical ways.

While we were driving homeward, Josh told me of a fond memory of fishing and friendship at the Farmington. He said, "Like many people, I found myself fishing a lot during the early days of the pandemic. That first summer in 2020 I fished hard—usually hitting the Farmington with my buddies Landon and Blake. We'd fish from dawn to dusk and even well into the night. That summer, we had one killer spot that we called 'The Campground' because it was near an actual public campground, but that is what made it feel so special. And the folks in the campground would start their fires and the smell of woodsmoke would mingle with the warm evening breeze as we cast our dry flies to rising trout. Man, that's something I'll never forget. It's a feeling I keep chasing . . . you know? Everything was so crazy back then, but standing in that cool river, it felt like the whole world just stopped for us, even if only for a moment."

As he told me this, I imagined it as if I were there with him, Landon, and Blake. And for a moment, I wished I had been. Perhaps one of the elements that make a place special is who we share it with. As a writer, I know my stories have little value until they are shared with others.

When I got back to the Inn at Stony Creek, Joe was waiting up for me as he had each night and poured me a celebratory glass of wine. We chatted about my time there, and he lamented that because I left so early each morning, he had never had the chance to cook me the breakfast that came with my stay, so we concluded that the next morning before Josh picked me up to head to the airport, I'd get my special breakfast. I slept well that night and may have dreamed of egg-and-cheese sandwiches and hot coffee, or short casts in fast water, or rising mayflies and falling spinners, or angry beavers at sunset—I can't recall. But what I do recall is the bittersweet feelings of saying goodbye to the Catskills and my friends Josh, Landon, Todd, Steve, and even Bill and Joe, the innkeepers. It's always good to go home, but I have found that everywhere I go I leave some of myself behind, or perhaps it is more accurate to say, I take something of those places and people with me.

It's a beautiful thing, this timeless connection. I was thinking of this as I sat outside on the slatestone steps of the inn, listening to the birds sing and noticing how the morning sunshine filtered through the trees. And that's when Josh pulled into the driveway to pick me up and bring me to the airport for my journey home to Texas. He stepped out of his SUV with a big smile on his face as if he was up to something—and he was. He said, "I saw how much you enjoyed Euro nymphing yesterday and I decided I wanted you to have this . . . it's the rod I started out with." I was stunned by his act of kindness and generosity and then by the irony of the moment. I looked at the name on the Moonshine Rod Company tube. It was called the Epiphany.

Chapter Eighteen

Wading the Battenkill, Vermont

If the only prayer you say throughout your life is "Thank You," then that will be enough.

—Elie Wiesel

Every day, I wake feeling grateful. I no longer pray to any deity, imagined or immortal. I don't cast my eyes upward while pleading for redemption or deliverance. I ask for nothing and expect nothing in return. I don't believe in or live out a transactional relationship with eternity. Instead, I simply choose to be kind and compassionate without expectation of reciprocity.

For me, the meaning of life comes from the meaningfulness I give to my life through my choices. I have no belief in my own importance to any omnipotent entity. The nature of the universe seems ambivalent to my existence—it does not love me, and I do not expect love from it. I do not strive to "follow the rules" but rather to simply live my life as best I can manage, with as much kindness I can give to others and this living planet. In my own way, I do pray each day by simply setting aside moments in time to be grateful. As I wake each morning and as I close my eyes each night, my first and my final words to the universe are always, "I am grateful." And I am.

When I go to the river, I seek nothing but solace, peace, and clarity. I do not count numbers of fish or value them by inches or pounds. I do not compete. It is not a game or a goal for me—it's a way of life. When I go to the river, I do not chastise myself for missteps—I learn from them. I laugh often and often I laugh at myself. And I don't allow myself to squander that solace, peace, and clarity by going to the river with those who do not likewise love and respect it. I fish and walk and live in Nature only with those who understand and appreciate the many magical and meaningful gifts that it offers the humble, grateful, openhearted soul. In short, life is too short to fish with muggles.

I'm no "expert" at anything and I never want to be one. I wish to maintain a beginner's mind where I am always learning. I currently know and use exactly three knots, none of which I tie with consistent competence. Some days my casting is in the groove, and other days it's in the trees. I have days where I can't seem to stop catching fish, and I have days where I catch nothing. Over the years, much of the dexterity in my hands and acuity in my eyes has diminished, and between my asthma and heart health issues, I can't climb up hills and over fallen timber with the ease that I once did. But all these challenges remind me that humility and empathy are as vital to my being as boldness and courage. In fishing as in life, forward casts and back casts reflect each other—and me.

When we arrived at the river, the crickets were singing their evening love songs. I mentioned to my friend Sarah how much I missed hearing crickets back home in Texas. There are things from my youth that I once thought to be ubiquitous, but now seem increasingly rare. They include evening crickets and lightning bugs and the lonesome morning calls of bobwhite quail. I also remember how midday rainstorms used to send choruses of frogs into synchronistic song, but now all too often the wetlands are dry, and the rains fall elsewhere. There are so many wildlife moments that haunt me because I experience them now only as memories. I miss the days of helping box turtles when they perilously chose to cross a country road in search of springtime love. Picking them up, I'd watch as they'd shyly enclose themselves within their brightly colored shells and then, once set down on the other side of the road, slowly open up while seeming to smile, as if grateful for my kindness. I miss these things that children of today might never know. And now while walking with Sarah and Brew toward the famed Battenkill River, I waxed nostalgic at the chirping of crickets, the buzzing of grasshoppers, and the way the wet grass brushed against my waders. It is within these breathless moments that true magic lives and breathes. It is in these vastly important tiny treasures that I find joy.

Sarah Foster is the executive director of the American Museum of Fly Fishing in Manchester, Vermont, just a stone's throw from the Orvis flagship store and the Battenkill River. I have the honor of being the museum's ambassador for Texas, and in that capacity I do what I can to support the good work of the museum in the preservation of and education about the history of American fly fishing. And I hope to have some small positive impact on the museum's ability to educate us all about the potential beneficial results of our hopefully evolving perspectives and sense of responsibility.

Sarah and the museum had invited me and other ambassadors to participate in their annual fly-fishing festival. I manned a booth where I met many wonderful people and sold and signed some of my books with the proceeds going to the museum. I also gave a brief presentation on the healing powers of Nature and fly fishing, which is an important theme in my life. If not for fly fishing, my love of Nature, and the love of my family, I would most likely not be here to write these words. Intimacy with Nature is healthy. Love of self and others is healing.

Brew Moscarello is a Battenkill River guide who is knowledgeable, patient, and kind. Sarah had introduced us to each other just moments before we reached the river, but it didn't take me any time at all to see that he was the perfect person to show us how to unlock a few of the secrets of the Battenkill. Before the evening was over, I knew he and I would become riverside friends. The thing I grew to admire most about Brew was his kindness and his support of Sarah. She is such a beautiful soul who is so determined to make a positive impact on her corner of the world. I'd fight zombies for her (though I hope I never have to prove that statement!).

Vermont, like most of New England, had been experiencing its own unique outcomes from increasingly rapid anthropogenic climate change that impacts everything from increased proliferation of disease-carrying ticks, to an increased northward flight of traditionally native saltwater forage fish and the gamefish that count on them, to in this case extreme rainfall and flooding, which in itself threatens aquatic life from the headwaters to the estuaries. The Battenkill was running unusually high, cold, and fast, just as I had experienced in my travels from Alaska, across the Rockies, and east to where I was standing in the Green Mountains of Vermont.

Prior to my arrival, people in the know, from Orvis's Tom Rosenbauer to almost everyone I encountered at the festival, said, "The Battenkill is tough. It's not unusual that a day on the Battenkill ends up being a day of casting practice in a beautiful setting." I was prepared for a potentially quiet evening with my friend Sarah in her home waters, but I could also see that Brew had a humble confidence about him that might prevail. I was going to enjoy the evening either way, but catching fish would be a wonderful bonus to this fishing trip.

A few months prior I had strung up a four-piece bamboo fly rod crafted by my friend Jerry Kustich of Sweetgrass Rods. As you might recall, Jerry graciously loaned the rod to me to try out while I was fishing in central Pennsylvania with my buddy Ross Purnell, the publisher/editor of *Fly Fisherman* magazine, for which I write the "Seasonable Angler" column. When I first opened the rod tube, I was struck by the monogram on the rod sock—"SR." Of course it stands for Sweetgrass Rods, but it sure seemed like it was made for Steve Ramirez when I started casting it and catching brown trout on the Little Juniata. Unfortunately, that was the same evening I managed to lose the rod tip on this extremely valuable work of functional art. Although I offered to pay for the entire rod, Jerry offered to simply build a new and improved rod tip and send it back to me in time for my trip to Vermont. Did I mention how gracious and generous a man Jerry Kustich is? I hope so.

I was eager to go retro with my borrowed bamboo rod, so I ditched my modern Orvis chest pack and decided to wear the original old cotton fishing vest that I started out with decades ago. It just felt right. We stood at the riverbank watching and waiting for a while, seeing what activity we might find and looking for bugs in the air or on the water. A few tiny caddis flies flitted above the surface of the river. They didn't last long, as flocks of cedar waxwings were dipping and diving from the overhanging tree branches and picking off every bug they could manage to catch.

I tied on a local version of a caddis dry fly that Brew gave me, but after watching the goings-on and lack of rising trout, he said, "Before you start fishing with the bamboo and the dry fly, I want to try something with my rod." I leaned Jerry's bamboo ever so carefully against a sheltering tree and began my first lesson in a technique I had never tried before but am so sold on now. We'll call it "caddis skipping."

Brew handed me an eleven-foot Orvis Euro nymphing rod that he said someone gave to him, but that he typically had no use for because he was not a fan of Euro nymphing, which he referred to as "not real fly fishing." Just a few weeks prior I had been standing on Connecticut's Farmington River with my buddies Josh Caldwell and Steve Hogan and had my first taste of Euro nymphing, and all I can say is, if it truly is the equivalent of a fly-fishing sin, I am a devoted sinner. For me it provided an intimate connection with the river's currents, substrate, and fish. But for Brew, Euro nymphing is unadulterated cheating and an unfair advantage. I can respect that. I asked questions about why he felt that way, and I understood where he was coming from, even if I did not become a convert. I was reminded of the claims I used to hear from dry-fly purists that nymph fishing with an indicator wasn't real fly fishing. I can understand that sentiment, too, without becoming a member of that church. I'm not religious in the least, and my spiritual soul feels just as comfortable and conflicted with one form of fishing or another. I was just glad Brew found a use for the wonderful long rod.

Since we knew this stretch of river held fish and good water to fish for them, but no fish were rising, the goal was to get those non-rising fish to rise. We planned to do this by imitating a caddis fly flying over the river and periodically touching the water to deposit eggs on the surface. The rig included a weighted nymph dropper under a caddis dry fly. The weighted nymph primarily acted to allow the dry fly to be manipulated so that it skipped and dipped across the surface. It's sort of a reverse hopper-and-dropper setup in that while it's possible to catch fish on the dropper, the goal is to catch them on the dry fly.

Brew demonstrated the technique, using the long rod to reach out over the current and then bob and bounce the delicate rod tip up and down so that the caddis dry fly seemed to dance across the surface. His demonstration was so effective that he immediately elicited a rise and take that led to him landing and releasing a nice Battenkill brown trout. He laughed and said, "I wasn't *trying* to catch one, but you can see that it works." Then he handed me the rod and said, "Here, it's your turn." Then he turned and waded downriver toward my sweet friend Sarah.

Looking downstream I could see her sending lovely soft casts across the river and into the currents. Sarah is a living and breathing smile, and watching her silhouetted in evening sunlight with a look of joy on her youthful face brought that same feeling to my aged heart. I meet the best people on the planet while standing in a river waving a stick. There's just something about this outdoor pastime that brings people of like mind and spirit together—and brings out the best in each of us. It's way past time for that kind of human community.

I could hear the soft murmurs of Sarah and Drew speaking and laughing together just downstream, but where I stood, I felt only solitude, which is not at all like being alone. One of the many magical aspects of fly fishing is that you can share moments in Nature with another member of your tribe yet do so in individual solitude in between those times of interactive friendship. I call this phenomenon "social solitude." You are never lonely, but always able to step away from the world and live in the moment with a peaceful sense of being connected to everything. And we are connected to everything.

The technique of caddis skipping took a bit of practice, but after a few floppy attempts and a couple of observations and recommendations called to me from upriver by Brew, I was getting the hang of it, and before long I had coaxed my first rise from a brownie that almost leapt from the river in its attempt to capture the dancing caddis. We both missed. I cast again.

There was an oval-shaped pool edged by fast current where that brown trout had risen, so I made a few more passes through there. What was possibly the same fish rose twice more without a hookup, then ceased to rise again. So I took a few careful steps upstream and across the slippery round-rock river bottom and cast again. I was starting to get the hang of the technique and really enjoyed trying my best to imitate a caddis fly depositing its progeny. After a few passes through the new run, a trout rose aggressively and took the fly. I set the hook, let out a whoop, and, after a short run and a jump, lost the fish. Merde. I tried again.

I continued to skip that little caddis through that same pool and received half a dozen strikes while losing several fish to poor hooksets on my part before finally landing a nice brown trout. Brew came up and gave me a few pointers on how to get a better hookset on a fly that you're bouncing off the water, and while he demonstrated this he managed to land another fish of solid size and ferocious fight. It was also gorgeously colored in sunset tones and shades. And it wasn't just the fish that looked that way.

After Brew handed me the rod and started back down toward my friend Sarah, I noticed that the sky was bleeding in yellow, amber, and golden swaths of pigment. The treetops seemed to be slightly ablaze, with their edges glowing like halos. For a moment, I simply stopped fishing and watched Sarah casting over a sunset backdrop and beneath the dimly illuminated trees and among the dipping and diving cedar waxwings. It was exactly how I might imagine any heaven to be—if I ever imagined such a mythological place. I don't. Heaven is wherever I'm standing.

With effort, I pulled my gaze from the setting sun and rising moon and turned back to casting and catching fish. On my first try, I got a rise but no take. After a few more passes I connected with another wild brown trout, this one a bit more mature than the last. I kept her in the water as I slipped out the barbless hook, admired her intense beauty ever so briefly, thanked her, and then sent her back home to the few feet of riverbed that was her eternity. I hope she lives a full and meaningful life that includes being born, growing up, thriving, reproducing, growing older, and then returning to the river—just like all of us.

I managed to catch a few fish using Brew's Euro nymphing rod that was never to be used for Euro nymphing. Then I returned it to him, hoping that Sarah could have a go of it while I invested some time casting a dry fly with Jerry's lovely Sweetgrass bamboo rod. Now I had to relearn how to cast slowly. After months of casting a fast graphite rod, I had to tell myself to "slow down."

There's something about bamboo that feels timeless. It's as if the back cast requires the caster to wait, breathe, and let the rod speak to you and tell you when it's time to cast forward. The rod is no longer a tool to be commanded. The angler becomes the extension of the rod, which is crafted not from bits of mined graphite but from the heart of a cultivated living being and the soul of the craftsman. It is organically grown, not inorganically extracted, and therefore contains an essence of residual life that runs from the community of soil and the energy of the sun to the stalks and leaves of the bamboo plant, and on through the hands, heart, and imagination of my friend Jerry. How cool is that?

It was beginning to get too dark to see anything, and we had already enjoyed a remarkable evening together casting to and catching fish on a historic fly-fishing river. As we reeled in our lines and spoke of the legendary nature of the Battenkill, I wondered about what makes a river "legendary" or "iconic" or "hallowed." Back in 1836 when Charles Orvis was born along its banks in the town of Manchester, was the Battenkill already considered legendary, and if so, why? It is beautiful, but no more so than many others I've fished. It is typical of these times that everyone who knows their home water describes it as "not what it used to be." Well, neither are we. There are now 8.5 billion of us still treating Nature as a "resource" and not as a treasured and intricately balanced community. And after traveling across this country many times now and fishing so many home waters, I've concluded that every water reflects its past and ours and is therefore "legendary."

But there's a bigger lesson to be learned here than simply recalling what once was by walking into a museum or wading into a river. Our past reflects our potential futures. Which shall it be? Will we live an abundant life, or exist with mere memories of it? As we walked back toward the trucks through the wet grass and the silhouettes of second-growth trees with the cheerful sounds of crickets filling the air, I knew what kind of life I wanted to live—and I was living it.

As I stowed my gear in Sarah's truck I could see Brew's outline in the darkness, surrounded by a circle of light coming from the cab of his pickup. Even in the dim light of dusk I could see Sarah's smile as she handed me an ice-cold can of Long Trail Brewing's Wild Blueberry beer, and we tapped our cans together and said, "Cheers!" And I was cheerful, except for a moment when it struck me that I was going to miss my friend when I returned to Texas the next day.

I felt so fortunate to have shared this evening with Sarah and Brew on the Battenkill, and I told her so. That's when she shyly said, "Thank you for accepting me." In that moment I realized that as executive director of the American Museum of Fly Fishing, she was used to being surrounded by "experts." I almost came to tears as I said, "Sarah, the honor and the pleasure is truly mine. I have fished with many, many people in many, many places and I can tell you that there's not one person I'd rather share this evening with than you." She smiled again and we hugged as Brew walked over and tapped his blueberry brew to ours—three friends sharing a moment under the stars. Three lives well lived.

If my description of this moment feels magical, it's because it was, and became even more so when I looked just past the shoulders of my dear friend to see flashes of light illuminating around her. Lightning bugs! I had not seen lightning bugs in decades. I felt like a child in his sixties. It felt like an instance of divine intervention—if I still believed in such things.

Perhaps I need to revisit and revitalize my faith in a universe that moves me in beautiful directions—when I'm open to being moved. Perhaps I am meeting everyone I am supposed to meet and learning what I need to learn. And perhaps dancing caddis flies and illuminating insects are there to remind me that there is always hope, if we live an actively hopeful life.

Living with hope calls us to heartfelt healing actions. It calls us to act in the creation of our own "good medicine" and the recognition of our innate magical abilities to help and to heal. I for one believe in the music of the spheres and the magic of the universe. No matter the challenge before us, love finds a way.

Perhaps the lesson that the universe is teaching me is that divine interventions happen every day in the smiles of friends and the flickers of light on dark, dark nights. As I stood there under the stars in the light of my friend's happiness, with the glow of life's exuberance shining all around us, I knew that life really is too short to fish with muggles. Every cast is hopeful. Every moment is magical. Every breath and every heartbeat are gifts that carry us around the next beautiful bend in the river of life. It's no wonder that I wake up each day feeling grateful. Don't you?

Part VI

Casting Homeward—The Foothills of the Ozarks and the Heart of Texas

Chapter Nineteen

Smallmouth Bass Fishing, Eastern Oklahoma

"The most important aspect of love is not in giving or the receiving: it's in the being."

—Ram Dass

I woke to rain falling ever so softly on the Ozark Mountains of eastern Oklahoma. I wondered if Emily was awake. After all the years of planning to get together, it seemed surreal that I was here at last, many months too late and yet right on time. Dave had "broke the tippet" one day before the previous Thanksgiving. We had been planning to get together and go bass fishing, but then COVID-19 came and time went by, and before we knew it, one of us had crossed the river, leaving the other two standing there pouring coffee and conversation into the empty spaces and noticing how the people we love never really leave us. They continue to live within us in both memory and action. Their soul seems to linger just beside us on a river or when we're lonely or while we sit, seemingly alone and listening to the songs of morning birds. They pass their kindness and wisdom on to us so that we may in turn give it to others. Every behavior can be learned and unlearned. Hate becomes a footnote that few ever read; love is a poem that outlives the poet. Love outlives us all.

When I walked out to the kitchen I could smell the glorious aroma of fresh coffee, and there was a mug sitting on the counter, preselected for me. I filled the cup, took my first grateful sip, and began searching for Emily. The rain had slowed to a light mist now, and I found her on the porch sitting in a chair that faced out across the lawn toward the trout stream that runs through the backyard. Next to her was an empty chair, and I hesitated because I knew this was Dave's chair. When she saw me she said, "Good morning," and invited me to sit there in the chair that contained so many memories of so many previous mornings. I sat down next to my friend, but wondered if I was worthy of such an honor. After all, this was Dave's chair, and I could still feel his presence. But I think he wanted me to sit there, because suddenly it felt natural and inviting, as if he were whispering in my ears, "I'm glad you finally made it up here, Steve. What took you so long?" And in my mind I replied, "I've been wrestling with that same question, Dave. What took me so long?"

Dave Whitlock impacted my life as he did so many others, and I never got the chance to share that with him as I wanted to—in person. As a young southern boy who grew up with cane poles and spinning rigs, I dreamed of fly fishing for my native bass. But as a boy, I was always told that fly fishing was something wealthy men did, and they did so exclusively with dry flies and only for trout. I'm so glad that Dave Whitlock knew better, and wrote about it, because it set me free to follow my own form of bliss. After all, I may have fished for delicately rising trout from Alaska to Montana and from Colorado to the Catskills of New York, but deep in my southern boy heart I am a bass angler. I think Dave was too.

The evening prior I had flown in on a flight from San Antonio via Houston where torrential downpours and lightning had grounded all air traffic for several hours and left me wondering if I was going to make it to Tulsa in time. I was scheduled to have dinner with Emily and attend the ceremony where the local Trout Unlimited chapter would be formally renamed the "Dave Whitlock Chapter." I was to be Emily's guest, an honor I took quite seriously. But the rainstorms left me feeling conflicted: I knew how desperately Texas and Oklahoma needed the rain, but I also selfishly wanted it to hold off just long enough to allow United Airlines to fly me through those unfriendly skies up to Tulsa where Emily was waiting for me. We had been discussing this trip for several years, and I needed to finally get there. And I did get there, too late for dinner but right on time for the ceremony. Then came the long nighttime drive into the Cherokee Nation where the Whitlock home resides and thrives almost eternally—like one of Dave's pencil drawings. Even in the dark of night, it felt like heaven.

We weren't in a rush that first morning because the air was crisp and cool, and a little time waiting for the rain to pass and the sun to shine would only increase our chances with the largemouth bass that awaited us in a nearby pond. When the rain stopped, we walked out to the bass and bluegill pond that Dave had designed, and at my request Emily gave me some pointers on improving my roll cast. I had been roll-casting in mediocre fashion for decades, but a few moments on the pond that morning with Emily did a lot to improve my technique and accuracy. And it struck me that Dave was there with us because he had taught these things to Emily, and now she passed them on to me. I am convinced that the value of our lives is measured in the ripples of positive impacts we may have on other living souls and on our one true home—planet Earth.

After our morning coffee and casting practice, we loaded up Emily's SUV and drove out to the ranch that contained one of Emily and Dave's favorite local bass ponds. When we arrived I could see that hidden behind its mundane edges were the memories of a million roll casts that came from the hands and hearts of Emily and Dave. I could hear the laughter and the cheers of mutual joy for each other. I could feel the peaceful, at-home feeling they must have felt together. And now that feeling was being passed on to me.

I could imagine the meaning this place had gathered beneath its waters and above its surrounding treetops for these two wonderful human beings who came here together through the years whenever they wanted to get away from "it all" and return to what truly matters. And it doesn't matter that this is a small and simple pond on a lovely horse ranch in eastern Oklahoma. These are hallowed waters. If a legend once lived there, are they not also legendary? Heaven exists in the eyes and heart of the beholder.

We found the pond's water level was so low that the canoe in the boathouse was inaccessible unless we slogged through knee-deep mud, so we opted for fishing from the far bank where the trees were fewer and the bass plentiful. I had brought my Orvis H-3 5-weight Jedi light saber, but Emily handed me one of Dave's rods with one of his favorite poppers tied on it, and then she gave me one of his fishing vests to wear. I didn't feel worthy, but she offered that honor, so I was honored to accept.

Inside the pockets I found boxes of Dave's hand-tied flies that he had created for Emily, each box decorated with his drawings of fish and dragonflies and containing loving notes to the woman who filled his life and heart for so many years. She had not changed a thing about the vest. It was exactly as he left it, complete with commentary about the contents of each box. I did not change a thing either, but wearing it may have changed me in that moment. I felt his presence. I felt his kindness. I felt their love. And now as I write this, I feel his presence once again—as if we are writing this together, and for me, this is such a pleasant thought and feeling.

Walking around the pond, we began casting over the muddy, sedge-covered banks and into darkish waters. Emily caught a couple of nice bass before I managed my first landing. That was no surprise, and I cheered for her each time, and she cheered for me whenever I managed to bring one to hand and then unceremoniously return it to the pond. The fishing was pleasant and peaceful. The catching was casual and calming. It felt like a lovely lazy day in the Ozarks being shared by two dear friends with all the time in the world—while wasting no time at all. We were living in the "now" while reliving a few precious memories of "then." We forgot about "someday." We felt timeless.

After returning to the house, we decided to do a bit more fishing, this time on the pond that Dave had designed and established behind the house. When Emily and Dave moved into the house, this area was nothing more than a depression in a big grassy yard. But Dave did the research about how to create a natural pond habitat and then set about constructing it here—in his backyard. Now it contained blooming white lily pads and tall green cattails from Texas and strategically submerged structure that led to the establishment of a thriving aquatic ecosystem where largemouth bass and bluegill shared the pond with ducks, songbirds, turtles, frogs, water snakes, dragonflies, and a host of other wildlife that could never have survived and thrived in a worthless grassy lawn.

One of the best things we can do to help Nature survive in our increasingly urban and suburban worlds is to make lawns a thing of the past and plant every "yard" exclusively in native trees, shrubs, wildflowers, and grasses. Providing a source of clean water is another thing we can all do to help wildlife coexist with 8.5 billion humans. As environmental writer Douglas W. Tallamy points out in his book titled *Nature's Best Hope*, small efforts made by many people can deliver "enormous physical, psychological, and environmental benefits to all" living creatures, including humanity.

At the pond's edge is the wooden dock from which we had practiced our casting earlier that morning. We cast off the canoe from that same dock later that evening. Also around the edges of the pond were several small casting platforms where Dave and Emily invested countless hours teaching children and adults the art and passion of fly fishing. Once again, our loving-kindness outlives us and passes from generation to generation, like spiritual DNA. Sometimes I wish that every child of Israel and Gaza would be taken fishing together, every formative year of their lives. If only we'd write a new story, not of religion or politics or generational hatred, but rather of friendship, understanding, respect, acceptance, and empathy. What kind of world might we create?

When Emily and Dave fished this pond together, they'd take turns paddling and casting. As her guest today, I was offered the bow as Emily guided us around the pond between the floating lily pad islands and into the many tiny cattail coves. I began catching bass almost immediately, with the occasional bluegill adding to the excitement here and there. If ever the day comes when I no longer love catching bluegill on a fly rod, I need to give my fly rods away. I hope that day never comes.

While we were fishing, Emily's mother, Betty, came out to one of the little casting platforms and began fishing for bluegill with a hook and bobber rig, and she started catching and releasing them pretty quickly. Well into her nineties, Betty is a bright force of nature who has a small home on the property. She enjoys each day riding around the grounds on her golf cart with her little chihuahua in the passenger seat as she tends her two vegetable gardens. She also loves to sit on the dock of the pond sipping sweet wine in a glass with ice chips while watching the ducks and cracking jokes that had us all laughing. After fishing the pond, Emily brought out some cheese, crackers, and wine, and the three of us watched the evening wane and simply enjoyed being alive together. For me, this is what fishing is all about. It's about being gratefully alive in every moment of every day. It's about connection, gratitude, and joy. It's about feeling at home on anyone's home water.

That evening we drank wine and enjoyed dinner while listening to the music of our era. We sang along to James Taylor, Carole King, Gordon Lightfoot, and the Eagles. We sat in chairs that were intentionally placed in front of a massive aquarium that Dave had constructed. And as we sat and sang and watched the fish swimming back and forth, I was keenly aware that once again, I was sitting in the empty chair that was full of memories. I've searched for a more descriptive word than "bittersweet," but can't find one. No other English word better describes my feelings of that moment. It felt bittersweet. It was a sorrow-edged joy—like a sunset, just as the stars begin to shine.

The aquarium was filled with wild native minnows and two crayfish, and we watched them swimming melodically back and forth as if they were dancing among the bubbles and the waving aquatic plants. Emily said that the aquarium originally had a few pairs of native sunfish in it, but they were too aggressive and ate all the minnows, so now only the minnows remain. Every living thing needs a safe place to live and thrive, and whenever we can help provide that, we are all the better for the experience. It was a restful night as the evening rains came once again, and in the morning, there was more coffee on the porch, more birdsong, more quiet and meaningful conversation, and ultimately, much more fishing.

The stream in front of the house used to be a "trickle," and perhaps long ago before it became a trickle, it was a thriving stream. This is "farm country" and every farm has a well, and . . . well, we know what too many wells can do to a water table. But just as he did when he designed and constructed the bass and bluegill pond, Dave designed and improved the stream habitat and now it contains a healthy population of rainbow trout—non-native, but naturalized, just like all of us. Now the stream has intermittent falls and riffles to help oxygenate the water, and deeper pools for the trout to feed and find shelter in from summertime predators and wintertime temperatures.

I told Emily that I wanted to catch and release a single fish in Dave's trout stream as part of my desire to engage with and honor his vision for this lovely landscape and the example he set for us all. But first we went to Dave's art studio where I sat for a while in silent solitude while Emily went downstairs to the tackle room to pull out a rod for me to use on the stream. And although Dave's studio is different than my friend Bob White's studio, I had much the same feeling—that I was in a hallowed place where spiritual alchemy unfolded. It felt like an empty church where divinity still lingers and the best of humanity silently surrounds you, while speaking volumes in its silence.

I sat at Dave's art desk and looked at his wide collection of colored pencils, paintbrushes, canvases, and reams of art paper, all surrounded by his yet-to-be-sold paintings and drawings on the walls and in racks throughout the room. I looked over at his fly-tying bench covered with the tools of the trade sitting as he had left them. And I perused his shelves of books and reread the words he placed on the back of the first book I ever wrote: *Casting Forward.* I felt his presence. I felt his genius. And I felt fortunate to have been his faraway friend.

Walking down to the tackle room I met up with Emily, and together we began walking along the stream and spotting a lot of big trout along the way. Besides the stream's aquatic features, Dave also ensured that the stream's riparian habitat was thriving with native trees, forbs, wildflowers, and grasses. Eastern red cedar, red maple, black cherry, sycamore, hackberry, river birch, elm, and burr oak lined the stream, along with winding stretches of wild grapes, wildflowers, and occasional patches of poison ivy. Even paradise has its perils.

When I saw the size of the trout in this small stretch of water, something bizarre happened to me: I became slightly unglued. First of all, Emily was giving me coaching pointers on presentation, and the more she helped me, the more nervous I seemed to get. To this day I have no idea what came over me, but somehow, I lost my mind and began acting like a newly minted graduate of a clown college.

To some degree I have lost my cool before when someone was watching me fish and coaching me as I tried to fish. It's my hang-up—nothing they are doing. Emily was being wonderful, but as she gave me real-time advice on my approach, cast, and presentation, all I could imagine was the memory of my fourth-grade math teacher standing there with a clipboard and a red-ink pen and shouting at me, "Nine times fourteen!" So I flubbed the cast and blew the presentation, and when a trout that seemed more like an orca finally pounced on the fly, I convulsed like a child who had a bee on him and totally blew the hookset! I knew what to do and had done it many, many, many times over the years, but with Emily watching me and offering advice as I cast to big fish in small water, I became a bumbling idiot who couldn't stop laughing at himself. It was tragically hilarious.

At one point, to demonstrate a technique, Emily took the rod and cast toward a part of the stream that she thought contained no fish. She immediately and accidentally hooked and landed a big trout. She seemed almost embarrassed at how easily she accomplished this, without even trying. So after releasing that fish she cast to another part of the water where she felt sure there were no fish, and she accidentally caught another one. I took the rod back, walked downstream a little, stopped laughing at myself, noticed Emily wasn't watching me anymore so cast the line, saw a fish rise, set the hook, and at last landed a fish. My moment of redemption was a relief, and I no longer felt as if I was wearing big floppy shoes and a red rubber nose. And yes, I am writing about it, because a writer should always be brave enough to say, "Occasionally, I suck at this." It's okay, sometimes I'm poetry in motion; this time, I was more of a limerick.

For me, fly fishing needs to consist of thoughtless motion and be meditative in nature. As soon as I start "thinking about it," I become a tree-snagging idiot. Whenever I relax and just let it flow, life is beautiful again and the fish seem to want to meet me. I want to meet them, too.

Although my plan was to catch just a single fish on this stream, once I caught the first one Emily suggested I try a few more casts just upstream, and when I did I connected with and landed another beautiful big rainbow trout. That was enough. I wasn't seriously fishing; that would come later. This was more of a ceremonial exercise so I could know the experience of fishing for and catching a fish from Dave's personally designed trout stream. I did, and it was done, so we had lunch beside the stream and laughed some more at my mini mental breakdown and ensuing recovery.

As much as I was enjoying everything we did on the pair of largemouth bass ponds and the one specially designed trout stream, the experience I was looking forward to most was yet to come. We were going to travel to and fish Dave and Emily's favorite little eastern Oklahoma smallmouth bass stream, one that I will name, "Love Creek." I gave it this name for two reasons. First, because sometimes we need to love a creek enough to protect it from being "loved" to death, and I promised not to write anything to give away its location. After all, it is semi-pristine because few anglers know of it and there is more pressure on the fish from otters, eagles, and herons than anglers—as it should be. The other reason was that, once again, I would be standing beside my dear friend Emily, in a place where the passing of Dave left an empty spot that I am no substitute for and can never fill. No one can. All the laughter and love these two wonderful human beings shared on these waters was so naturally obvious to me, and I felt that joy lingering over this water like evening sunlight. So it seemed natural to me that a stream surrounded by so many memories of love, one that so desperately needs to be loved, ought to be called "Love Creek."

Love Creek runs through a large private ranch where horses and cattle roam the uplands and wildflowers, wildlife, and native wild fish inhabit the riverine habitats. As we pulled into the ranch, Emily asked if I minded opening and closing the gates, to which I replied, "I'm a Texan . . . it comes second nature." And it does. I grew up working on my best friend's ranch chasing cattle and feeding horses. By the time I was eighteen, I had someone else opening the gates for me because I was riding the rodeo circuit while strapped to the back of a bull in eight-second intervals, each feeling like an eternity. For me, crossing ranch country and opening gates is like coming home.

Once we got through the last gate, around the ancient walnut trees, and past the pastures that once were prairies, we came to a large, flat river bottom that was covered in native grasses and wildflowers. We'd have to cross through this knee- to thigh-high brush for about one hundred yards before we'd reach the stream, and from past experience my instincts told me that it all looked sort of "snaky," and I said as much as we waded into the meadow.

I should mention that I grew up in snake country and love and value snakes as magnificent and important parts of any ecosystem they naturally inhabit. I lament their rapid depletion from habitat loss, pickup truck tires, and purposeful human predation. Snake and human interactions that go south are more times than not due to the poor judgment choices of *Homo sapiens*—"wise man."

All snakes are short, about two inches tall. To them, we humans are scary giants. With that said, I know to respect these beautiful creatures and give them space so they do not feel threatened. I suggested to Emily that we cross through the area of least brush, where we could best see exactly what we might be stepping on or near. I asked her if she'd seen any western diamondback rattlesnakes in the area, and she said that she had not, and that although they have copperheads and western cottonmouths in the region, she had not encountered either along this stream. My approach was still cautious and aware, which is the result of decades of near misses with death—instant or lingering.

When we entered Love Creek, it was everything I had imagined and more, with the banks lined in chest-high purple perilla flowers that were covered in Ozark color and European honeybees. A mature bald eagle flew upstream along the tree line and out of sight, and a great white egret angled just downstream while keeping a wary eye on us primate intruders. Millennia of encounters with humans have been embedded in the genetic code of every wild creature. We've earned a bad name in almost every corner of the planet. Even the crows and ravens have devised ways to warn each other that "the humans are coming." I'm guessing that mastodons and mammoths might be around today had they been a bit more talkative and quicker on their feet. We'll never know.

The waters of Love Creek sustain an ample mixture of forage and game fish. Along the river bottom skirt the bright little bodies of sculpins, darters, suckers, crayfish, and various aquatic insects. Mid-water forage fish include a plethora of native minnows, which add to the frogs, tadpoles, and various terrestrials that are present for the larger fish to feed upon. The larger fish include northern largemouth, northern smallmouth, Kentucky spotted, and the endemic and genetically unique Neosho smallmouth bass. Other warmwater species, including bluegill, pumpkinseed, and green sunfish, fill the final fishy niche of Love Creek. And all of this fits in a stream that is most often no more than thirty feet wide and three feet deep. It truly is a paradise.

As I write this, I am sitting at my desk at home listening to Tim O'Brien and Darrell Scott singing the classic John Prine song "Paradise." This was one of the first songs I ever learned to play on the guitar, and I will never forget that day so long ago when I sang of paradise in the presence of my stepfather Paul Church. Paul was a good man who died half a dozen years ago from cancer but who made his living selling cranes and other heavy equipment around the world. The song tells the story of a young man from western Kentucky asking his "daddy" to take him home to the rivers and mountains where he grew up, which he describes as being "paradise." The father tells the young man that he is too late because "Mr. Peabody's" coal trains have hauled his home away. When I finished singing the song, I noticed my stepfather had become quiet and thoughtful. I asked if everything was all right, to which he said, "That song made me feel sadness and regret. I once sold a lot of heavy equipment to Mr. Peabody." We never know the everlasting ripple effects of our actions and inactions. Once something is gone, it's often gone for all time.

I think about that whenever I discover a place like Love Creek. These national and ecological treasures are all just one coal or copper mine, gravel pit, roadway, factory or corporate farm discharge, or water extraction project away from extinction. And once we've turned our back on these magical places, there's no turning back from the outcome of our ambivalence.

As I stepped into the stream, I felt as if I was stepping into hallowed waters—and I was. I looked up into the trees and searched to see where the eagle had gone but could not find him. Still, I knew he was up there, somewhere. And it is the continued existence of living beings such as bald eagles and great white egrets that gives me hope. After all, not long ago we had pushed both species to the brink of extinction, and now they are coming back. There is hope and the hope resides in each of us—together. In spite of the many signs to the contrary, I remain hopeful.

We began taking turns casting into a long pool that had a small waterfall emptying into it, plenty of tree roots and grasses along its undercut banks, and a riffle of well-oxygenated water both up- and downstream of it. We could see several bass hanging in the current just under the submerged tree roots and took turns casting to them. Emily caught two fish for my one, but we both pulled a few out, including northern largemouth and smallmouth, plus a few green sunfish, and Emily landed a Neosho smallmouth. It was a good spot to say the least, both for its raw beauty and its abundant and eager fish.

Walking though the shallow riffle, I began casting toward an undercut bend in the stream while Emily leapfrogged down to the next pool. I began sight casting to a number of good-size bass and got a few follows but no takes. I had switched to a Woolly Bugger at the end of the last pool, but now decided to switch back to the popper, as Emily was catching quite a few on top water and I wasn't finding any takers, which seemed unusual for bass fishing. I was surprised that these fish were so selective, but there was so much food in this little creek that they didn't have to work too hard to fulfill their ravenous desires.

After switching back to a topwater popper fly, Emily invited me to work the pool where she had been quite successful, while she tried the next one downstream. Once I waded down there, I could see what a wonderful bend in the stream it was and why. There was a long, deep pool with ample current and an undercut bank of rocks and roots. Love Creek is fed by clear, clean, limestone spring–fed waters, quite like those of my beloved Texas Hill Country. The fish had everything they needed to thrive—water quality, sheltering habitat, abundant food sources, and most of all, limited human interaction. It didn't take long for me to begin connecting with fish, and I was surprised to see the plethora of species that I was encountering. In a short span of time and space I landed northern largemouth and northern smallmouth, Kentucky spotted bass, green sunfish, bluegill, and a single stunningly beautiful red-eyed Neosho smallmouth bass!

Emily and I worked our way downstream and caught a few fish here and there while also taking time to just be present in the moment together, listening to birds, watching deer move silently through the understory of the forest, and feeling the gentle push and pull of Love Creek and of life itself. I'm convinced that life is more than the flicker of our brief biological existence. I intuitively feel the living consciousness that flows through me and Emily and the fish I hold in my wet hand, and the limestone stream that holds me. And I so wish we temporary humans would all wake from our daydream existence and realize that the petty illusions we treat as realities are not worth a single cast or consideration.

Can you image a world where we simply lived like a spring-fed stream? What if we gave life to everything we touched and cycled ourselves willingly from one form to another? We are traveling that path anyway. One day, I will be the ocean again, and so will you. Just like Dave, we all eventually break the tippet. It comes naturally for us transitory beings made of stardust and raindrops. Ashes to ashes, dust to dust—the same river flows through us, one and all.

Emily and I walked back upstream to the first stretch of water where we had begun the day, the same place where we had carefully walked in through the thigh-high wildflowers and grasses. We both knew these would be our final casts as the sun slipped over the treetops and the air grew cooler. And we were content with this knowledge, because we had shared a magnificent evening together on this precious ribbon of life and living. And besides, there was good food, wine, music, and conversation waiting for us back at the house. At the end of any day, I'm fishing for experiences—not fish. There is no use in keeping score. As Papa Hemingway once observed, the winner takes nothing.

It was one of those soft, amber evenings where the light catches the grasses and leaves sideways, and everything is edged in fragments of that life-giving light. The riffles shimmered like bits of broken sea glass reflecting the colors of the forest and fields. We were both alone in our thoughts as Emily cast just upstream of me, and I cast to and caught a few final smallmouth bass. I could still hear the honeybees buzzing in the purpleperilla flowers, and a barred owl began calling from somewhere just inside the streamside forest. I was casting, stripping in line, moving with sideways steps downstream, and casting again in a slow and methodical fashion when I suddenly felt as if someone had whispered in my ear, "Look before you step, Steve." And I did. And when I did, I saw one of the most beautiful water moccasins I have ever seen, coiled up in a defensive stance, floating on the water just downstream of my right leg. When I saw her I said, "Well hello there! Aren't you a healthy-looking girl?" Then I told Emily what I had discovered, and she began to wade downstream toward me to take a look.

I didn't want to hurt this pretty but potentially dangerous creature, but I needed to get it to move so we could get to the part of the embankment where it was safest to walk back toward the vehicle—after all, there might be more venomous snakes here. Emily said, "Just toss a rock near it so that it will move out of our path." I bent down and picked up a river stone about the size of a Ping-Pong ball and tossed it into the water beside the cottonmouth. The snake coiled even tighter, but remained fixated on my seemingly threatening figure, her forked tongue waving slowly as she tried to pick up every bit of scent so that the question "What is this creature that is throwing rocks at me?" might be answered. I picked up another stone about the size of a tennis ball and tossed it next to the snake; it made a plunk but provoked no reaction other than increased vigilance and alarm from the hapless serpent. After all, she was only trying to go fishing in solitude when I walked up to her—even if she did "low hole" me. That's when a boulder about the size of a football came flying over my shoulder, creating a tsunami-like wave that sent the serpent fleeing quickly toward the next county. Once again, Emily showed me how it's done. So we crossed the meadow, loaded our gear, and headed for home. On the way out, I got the gates.

I guess home is wherever our heart is, and as we sat together watching the aquarium fish swimming while sipping wine and singing along to the music of our youth, my heart felt full. I began thinking of how we had lost both John Prine and Dave Whitlock in this time of epidemic, war, political unrest, and ecological collapse—all caused primarily in the name of profit and power. I was recalling how John sang of his "Paradise" flowing along the Green River of Kentucky, and how in keeping with his wishes some of his ashes were set free into its currents. And I thought of the glass vial of Dave's ashes that would travel with me to my Texas Hill Country home waters. Emily had asked me to return part of Dave to the Devils River, a place they had both grown to know and love. To do so is my honor.

Author Terry Pratchett once wrote, "No one is actually dead until the ripples they cause in the world die away." I guess that's why we once painted on the walls of caves and carved the hearts of trees and bones of whales, and why I write fishing stories that have almost nothing to do with fishing. Those of us who paint, carve, and write know that we leave nothing behind but the memories, moments, and meanings that we give to our lives through the loving kindness and courage we embody in our lives. No pharaoh has ever returned for his things; they remain in tombs and museums, as they should. Our legacy is not within our ashes but rather from the light of our fire within. If the results of our choices are what we leave behind, perhaps this is also what determines what we take with us.

When I got home, it was raining. I did not feel sorrow; I felt gratitude and joy. The rivers need the rain, and I need the rivers. After all, I'm hoping to make a few more healing ripples while my life still flows.

Chapter Twenty

Floating the Llano River, Texas Hill Country

Some people talk to animals. Not many listen though. That's the problem.

—A. A. Milne, *Winnie-the-Pooh*

One of the aspects of any adventure that I love is we never know what we will learn along the way about the universe and our place in the universe. Teachers and lessons come in many forms. Sometimes they are the fish we connect with, or the ones we never come to know. Sometimes they are the people we are fishing with and the way they reflect our own journey, or contrast with it. Sometimes lessons are hidden in the songs of birds or the way the evening sunlight touches and transforms the landscape. But in this case, my teacher was the river itself with its winding ways and the push and pull of its circuitous currents. And although fish were caught and returned to their home waters, this fishing trip was not about the fish and even the fishing—it was about the river and the lessons it teaches about life and living.

The Llano River is born among the limestone springs of the far western Texas Hill Country. It meanders across the Llano Uplift where pink granite, feldspar, and quartz of volcanic origin prevail, and where red-winged blackbirds sing from the cattails and fence posts. This place, like all places on planet Earth, is both wild and feral. It contains hints of what once was before the arrival of humans, and glimpses of the ghosts of living beings that have vanished since that time—both human and not. Spanish explorers seeking cities of gold once built walls of stone around their Catholic missions and well-armed military compounds. German immigrants seeking to establish freethinking utopian societies once built fences of juniper and barbed wire, but the results were much the same. The walls and fences crumbled, and the golden utopian communities faded as capitalism outbid compassion and religion overruled reason. But whenever I have drifted down this living river, the memories and mementos of past lives who once called this place home remain to remind me that there truly is gold in these hills. It comes in the form of sunsets and autumn leaves and the way the evening light glows among the bending grass and winding waters. It comes in the way the birds sing and the people smile and wave as you drive by. It lives in the tin roofs and limestone walls of farmhouses that blend with the surrounding woodlands, and the white-tailed deer whose ancestors once browsed among herds of bison and kept a wary eye to Mexican wolves, mountain lions, and jaguars. As my dear friend Cari Ray and I launched her raft into the Llano River, all I could think of was the word "Home."

I don't know why I chose to bring my trout rod and line on a bass fishing trip, but I did. That's like bringing a butter knife to a gunfight. But after a few casts and a quick catch and return of a pretty Guadalupe bass, we both decided that the big bass flies we were tossing were a lot more cooperative when I used the heavier and slower glass rod that Cari loaned me. It was an early 1970s version Fenwick FF806, an eight-foot-long glass 6-weight that she bought in a thrift store for twenty-five dollars, and it was paired up with a heavy line and shooting head that was specifically designed for quick casts with clunky bass flies. Let me tell you, I was in love, or as Cari said, I was "digging it!"

Just like I was connecting with that old Fenwick glass rod, there are a rare few people that I meet in life with whom I feel instantly at home. Cari Ray is one such person, who like me is old of soul and forever young at heart. Even with over a decade of distance in our biological ages, we travel much the same winding and uncertain path. And no matter if it's across an urban table eating tacos or on a raft drifting down a wild river, we can share almost anything and feel safe, respected, cared for, and understood. I guess that's another way to feel at home, not with a place or even a perspective, but rather with a person. Cari and I are homies.

The river was recovering from perhaps the greatest drought in recorded history, where entire sections of the Llano, Guadalupe, Blanco, Pedernales, and other Texas Hill Country rivers had gone completely dry to the bare limestone bottoms, with every bit of aquatic life either stone-dead or suffering in the rare few oxygen-starved puddles. With the anthropogenic climate change, each summer becomes more unbearable, with months of consistent triple-digit temperatures now the new normal. And with San Antonio and Austin growing exponentially in their populations of thirsty humans and worthless lawns, the aquifers are being drained almost beyond repair. It's like watching someone you deeply love as they desperately cling to life while the light dims in their once bright eyes.

When Cari and I launched the raft and saw that the river was once again flowing due to recent rains, we felt relieved to see our dear friend "breathing" once again. I will hold this river's hand and stay by her side, to the last drop of water. She matters. Communities of living beings depend on her. I depend on her.

Cari was on the sticks, and I was casting that mystical twenty-five-dollar fly rod as if we were meant for each other—so much so that Cari joked she was getting worried I might never give it back. She need not worry—I'm not a homewrecker. But I was having fun dropping big, clunky bass bugs wherever I wanted them to go, and I managed to land a few more Guadalupes, a decent-size largemouth bass, and a couple of chunky sunfish along the way. I wanted Cari to have a chance to fish and offered to row, but she dropped anchor instead, and we both began casting and catching in likely spots around the boat. It was a perfect day in the best of company. Still, I'm not sure Cari Ray quite trusted me to row her boat—which was probably wise. Perhaps I have a reckless gleam in my wild-man eyes. I sure hope so.

The float we had planned was about five miles in length and timeless in the way it transported us along a truly wild river. So far, the Llano remains without dams or diversions for most of its journey, only accumulating a few such scars as it grows closer to the Colorado River and the "civilization" that has become the city of Austin. It's no secret that the upper Llano River is one of the places in my beloved Texas Hill Country that I'd like to see designated as a National Wild and Scenic River. The other is the Devils River on the far western edge of the Hill Country. If I can help this dream become a reality, I will.

I was casting and catching as Cari rowed again, and we reminisced about our many memories of feeling truly alive on the Llano. Rivers are like that. They remind you of the moments that were filled with birdsong and the plop, plop, plop of turtles sliding off logs or the way the evening sunlight shimmered as a plain-bellied water snake swam effortlessly across the current. They cause your mind to flash back to images of laughter and love, and the time it was too cold to fish so you sat on a low bridge with a dear friend drinking wine and listening to the sound of the cold, clear water as it tumbled over golden limestone or soft pink granite.

After a while of my casting and catching and Cari rowing and reminiscing, we landed the boat on a large granite island and fished together—each cheering the other as we brought bass and sunfish to wet hands and then watched as they swam away free—taking us with them. And that's about the time that Cari pointed to a floating tree limb that had all of its bark chewed off and said, "Look! It's beaver wood!" And I said, "I've been walking these shorelines and fishing these rivers for decades, and although I know we do have beavers here, I've yet to ever see one." Cari and I have both seen many beavers in the wild, just not in our Texas home waters. Rivers are like that; they leave you wondering what comes next—just like life.

And that's how it was as we launched once again into the currents, with Cari rowing not so much to propel us forward as to keep us in line with those living and ever-changing currents. At regular intervals the river would divide into several possible forward pathways. In each case we could not see what waited for us around each bend. So we would hold in place long enough to watch the way the river moved through each potential passage. We'd listen for the sounds that the river might make just around each bend. And we'd take notice of overhanging trees and submerged rocks and roots, and braided shallows where we might have to get out and drag the boat. The only way to discern which way to go was to pay attention to what the river was teaching us. But in the end, you make your best guess and adapt to whatever reality awaits you—again, just like life.

The sky was that uniquely western shade of blue with just enough cotton-white wispy clouds to give it character. The life-giving light of our lone star framed each mesquite tree in a golden glow and caused the bending bunchgrasses to burn bronze and copper in the breeze. We watched breathlessly as three white-tailed does walked silently through the evening light along the river's life-giving edges. The water of the Llano had the special translucent emerald-green quality that Cari observed as reflecting the sky and the trees and even the pebbled riverbed below. Dragonflies of many shapes and colors darted and rested on the bunchgrass and mesquite. We could not have imagined a more magical day—and yet, we were living it.

In time we came to a deep and dark backwater eddy, and Cari cast into it with something brown and buggy on the end of her leader. After allowing ample time for the fly to sink into the best potential feeding zone, she made a short strip and instantly connected with something heavy and brutish. Her Epic 476 "Holy Ghost" bent toward the water with such earnestness that I began to wonder who had caught whom—friend or fish. "This could be the fish of the day!" I said. "I wonder what it is," she replied. And at that moment I knew we both had visions of a massive largemouth bass with a gullet that could swallow a kayak and a look of malice in his eyes. But instead we saw the culprit and proclaimed in one disappointed voice, "Catfish."

It seems that another thing Cari and I have in common is that neither of us enjoys catching catfish on the fly. If I were a more perfect Buddha, I would value everything the same—and I do value these fish and their place in the ecosystem or a frying pan. But they are a pain to get off the hook with their rubbery mouths, and more than once I have been stabbed by their dorsal fins, so yes, I'd rather not catch them. I'm also not a big fan of cooked carrots. I'm an Imperfect Texan Buddha indeed.

We came to another fork in the river where we needed to decide which path to choose. Cari dropped anchor, and we looked at the left and right passages and considered what the river and the riverine shoreline had to teach us. We listened for the sound of rapids but couldn't tell where it was coming from—left, right, or both. We took notice of the vegetation and the amount of submerged and overhead obstructions we could see or discern. And we paid attention to the speed of the water flowing into each river path and the locations of gravel bars and undercut embankments. But in the end, it came down to our best guess, and we chose to row into the left branch and around the corner through the currents and eddies until we found what awaited us around our chosen corner. It was a waterfall, and there was no turning back. This too is much like life.

Sometimes no matter what we do, we are given a challenge. It could be asthma, heart disease, or cancer. It could be night terrors and the lingering anxiety of post-traumatic stress. It could be a crumbling of community or the erosion of collective empathy. It could be the potential loss of our democracy or the collapse of our environment. But whatever we face, we can face it better when we do so together and with an open mind and willing heart. And that is exactly what we did. We landed the boat and guided it by working together with bow and stern ropes from the safety of the shoreline, and in the end our little vessel made it over the series of cascades and we continued our journey down this living river—together. Inside every challenge there is a gift in the form of a lesson and an opportunity to grow and thrive. I am not merely a survivor; I am a thriver. How about you?

The next stretch of river was gentle with us, and we drifted and cast toward likely banks, plunge pools, and runs, picking up a bass or sunfish here and there while enjoying both conversation and silence. Cari and I often seek riverside wisdom when we fish together. We speak of life and death and everything in between. We speak of the people we love and who love us, and of those who for whatever reason had vanished. We share our impressions about the difficulties and opportunities in life's many transitions and unexpected outcomes. And we considered how brave it is to give your heart to a person or a river knowing that one might drift away and the other might dry up—and there's little we can do about either eventuality. Love is the most painful and powerful thing I know—and the most beautiful. I will never forgo love in a foolish attempt to avoid loss. Love and courage are symbiotic. You can't have one without the other. Whenever I meet a loveless soul, I know they are fearful.

Every journey has its beginning and its ending, although exactly where these moments reside is up for interpretation. I believe everything begins in our imagination and ends only when we stop living it and learning from it inside that same yet ever-changing mind. So in that sense I am still drifting down the Llano even as I sit here at my desk writing this story so that I might share it with you. But in temporal terms, Cari and I were coming to the end of our float. We had arrived at our final choice as to which path was our true path.

The river split into three distinct pathways, and in each case due to the topography and vegetation we could not see what awaited us around each turn. So we paused and considered the things we could control, then went with our gut and chose the right turn. As Cari began to row us into the current and around the bend, we braced for whatever lay ahead. True adventure is about adapting to the unknown once it becomes known, and this was a true adventure.

Just around the bend we came to a fast-moving narrow with overhanging mesquite trees on one side and shoulder-high bunchgrasses on the other. We could see where the river was opening up just ahead, and so we held on tight as the currents and Cari moved us under the tree limbs and over the submerged rocks and roots. That was about the time the magic happened. A puff of gravel and sand exploded into the streamside currents, and at the end of that cloud of debris was the largest beaver I had ever seen and the first one I've ever met in my home state of Texas. I was beyond thrilled to see him, and we seemed to be communicating with each other as we made eye contact and recognized each other's aliveness. At one point the beaver crawled up on the riverbank and just sat there with not even a rod's length between us—me looking into his eyes and him looking into mine. It was obvious that we were connecting on some primal level, and in the connection we felt at home with each other.

For a while he swam just in front of us as we cast our line in the hope of catching and releasing a few more fish before the takeout. But I really didn't need to catch any more fish. We had chosen wisely, and the currents of the river or the kindness of the universe had given me the gift of this brief but poignant introduction. In that moment two lives sharing the river became three. A chance encounter or a passing word can change the direction of a lifetime. We never truly know the ripple effect of our choices. Life and learning are intertwined. Without growth, there is only death. I for one want to live—truly.

After we had loaded up the raft on Cari's trailer, we began to drive westward toward the sleepy sunlight and the waking starlight. Cari pointed to the horizon and said, "Look at that magnificent pink sky!" I'm pink color blind, so what she saw as pink I saw as other shades of painted skylight and fading colors. For me, the sky was blue fading into purple and orange, yellow, amber, and perhaps a salmon-like red where she saw soft pink. But we both saw our own reflected impressions of beauty and magic.

We must all run our own race, envision our own dreams, and cast our own lines forever forward. And if we do so together with the same core values of gratitude, kindness, empathy, respect, and childlike curiosity, we will create beauty—first in our imaginations and then in our realities. Love and courage conquer any challenging current we may encounter. And that, my friends, are what this river of life has to teach us. Our perspective is reflective of our choices and our choices are reflective of our chosen perspective. It's all connected. It's all a circle. It's how we live a life worth living. No matter what comes around any bend, just keep rowing.

Chapter Twenty-One

Just One More Cast—Devils River, Texas Hill Country

I'm not saying that everything is survivable. Just that everything except the last thing is.

—John Green, *Paper Towns*

The Devils River is alive with spring-fed waters that are so transparent that a raft or kayak might seem to be floating on liquid oxygen as it glides over a furled limestone floor and under a cobalt-blue ceiling. Its walls are constructed of ocher- and onyx-stained limestone bluffs—remnants of an ancient seabed where aquatic dinosaurs once hunted the ancestors of the modern fish we were seeking. In summer the temperatures can be life threatening, while winter temperatures can dip from pleasant to below freezing in a matter of hours. Here, a person might experience all four seasons in a single day. It's a harsh and magnificent landscape that is transformed by an unlikely ribbon of water cutting through its dry, thorn-strewn, and stony heart. The Devils River is a riparian heaven.

I have traveled much of the world and explored many rivers, but for me, the Devils River of Texas is my most sacred place. For me the Devils River contains holy water that miraculously transforms a parched land into a paradise. I'm so grateful that Cari and I decided to travel across the Edwards Plateau so that we might share a few days and nights together in this precarious paradise. It just felt right. It also felt urgent. It felt like a story that must be told.

When we launched Cari's raft, we did so with the intention of taking a five-mile float from the "Devils Back" launch site to our encampment some two miles below Pafford Crossing and about ten miles above Rough Canyon, a place described as having consistently "strong headwinds and choppy water." Where we were fishing was not far from the Rio Grande, or as our southern neighbors might call it, the Rio Bravo. By any name it has become a waterline of division rather than a place of common ground. For countless generations families have worked and lived on both sides of the river, and as is the case with most of humanity, the vast majority of these people lived and still live in peace, if not prosperity. Last night we diced some tomato and avocado into our fireside soup. On each fruit was a sticker that read, "Product of Mexico."

Above us were high limestone bluffs with a massive eagle's nest inside a rock shelter and Chihuahuan Desert ocotillo plants pointing like fingers toward the few white clouds that were suspended in the bluest sky I've ever seen. Along the river the sycamore trees were turning golden yellow with the recent cooling nights, although autumn seems to be arriving later each year and winter becomes more of a nap than a long night's sleep.

The night prior Cari and I sat by a roaring campfire until almost midnight sharing thoughts about life, living, and of growing older and increasingly introspective. And after, we endured a night of freezing temperatures that kept us awake despite our many layers and heat-gathering sleeping bags. December is not prime fishing time on the Devils River. Warmwater fish get moody in cold-water times. Cari rowed us to a limestone island in the river that was covered in a few inches of water, and we both got out and began casting our lines toward the likely lairs of brooding bass.

Cari started casting downriver along a line of submerged limestone shelves while I did the same just upcurrent from where she stood. The sun was out, and I was feeling hopeful of some topwater action. On about my third cast I received a slashing pass where I could see the dorsal-finned back of the bass that swiped at and missed my popper. It felt like a good sign, and so we kept working the area methodically, but to no avail.

We loaded back into the raft and crossed the river, which here at the Devils Back was wide, vast, and without much current. To complicate things we had a powerful headwind coming up from Mexico. Once across we landed the boat on a limestone ridge to reconsider our plans. Cari was a bit concerned about how hard of a float it might be with that gusting wind in our faces and her buoyant little boat acting as a sail. She had reason for concern. After all, we had no idea what lay ahead, but every notion of what surrounded us in this place and time.

As we considered our options, I was reminded of a conversation with my dear friend Sue Kerver where she recounted how happy she was with her life, but that she felt the pressures of societal expectations to keep seeking "advancement" even if that journey meant leaving the life she loved. After listening I had said, "It's rarely wise to leave fish in an effort to find fish." As I stood below the majestic cliffs and above the magical waters of the Devils, it occurred to me that our choice was much the same. You should never leave paradise in an effort to find paradise. I suspect the real lesson of the biblical characters "Adam" and "Eve" is that when you find yourself in a beautiful place, don't screw it up. With this lesson in mind, we decided to forgo the long windswept float and undergo an in-depth exploration of the paradise that flowed all around us. It was a wise choice.

We began exploring along a massive limestone shelf that ran for about a half mile along the far side of the river, with Cari using the raft to fish the southbound edges and me using my wader-clad feet to explore upriver. It didn't take long for Cari to connect with and then lose a fish of indeterminate species and then immediately redeem herself by catching the first smallmouth bass of the day. I remained without redemption.

I was happy for my friend because it was her first Devils River smallmouth and I had been here and caught them before. But after she caught another and then another and I was still standing there alone and unloved by a single fish, I decided it was time to give up my topwater dreams and tie on an olive streamer like my friend. In retrospect, I should have tied on something completely different, so that if I remained unloved and without redemption, I could wonder if I simply needed to change flies rather than be certain that it was all my fault.

I began walking upriver and casting out as far as I could along the foam lines, clumps of half-submerged bunchgrass and deep cuts in the limestone bottom. I could see a few bass way down deep, including a few big largemouth and some moderate-size smallies, but they were ignoring my offering. So I cast out toward a limestone ridge that had shaded water along one side and immediately received a strike and hookup! It was a chunky sunfish—but not the bass I was searching for. I cast again and caught another fish, and then another, and then another, each of the half dozen or so fish I caught and released slightly bigger than the last, but all were sunfish. I debated switching my fly yet again. But instead, I stopped fishing for a while and took a few moments to pick up my head and simply absorb where I was and what being there in that moment meant to me. I'm so glad that I did.

I'm usually pretty adept at not taking my angling too seriously. I mean, think about it. I travel long distances so that I can catch fish that I immediately release. But the truth is, I didn't come here for the fishing as much as the fishing was just a reason to come here and immerse myself in this place and time—literally. I came here to experience the healing powers and poignancy of Nature. I came here to breathe the clean air and feel the warming sunlight and cooling breezes. And I came here to seek answers to the questions that have been raised by my aging and ailing body, our darkly changing world and collapsing environment, and my growing sense of urgency to do what little good I can manage while I'm still here.

I get the feeling that my late Marine Corps brother Dave was right when he came to me in a dream and said, "Whatever you feel you need to do in this lifetime, you need to get on with it. Soon none of this will be here, unless you do." It's hard for me to believe that my buddy Dave De la Garza passed from this world some twenty-one years ago. And it's hard for me to grasp that my other Marine Corps brother Monty Lambert has been gone for almost three years. And now, it's just me here, standing in this beautiful river that, like me, must fight for its life.

Just as my evolving asthma and heart disease reduce the flow of life-sustaining oxygen in my body, the many legal and illegal wells that are poised to drain the Devils River threaten its continued existence and every living thing that is sustained by this spring-fed miracle of Nature. But this is a message of hope as an action verb because, just as I have been able to improve my own health through lifestyle changes of diet, exercise, and medical care, we can change the trajectory of our current social and environmental life expectancy from declining to reviving. It's all within our power. Everything except the last thing is survivable.

The Devils River is not a "resource" to be used or a "challenge" to be conquered. It is a national and natural treasure to be respected, valued, and loved, but never loved to death. Through my words and actions, I am trying my best to be a fisher of men and women, who is constantly striving to learn with humility and teach with a sense of service and a commitment to courage. And as I watched the sycamore trees dancing in the wind and the clear, cold water advancing toward the sea, it didn't matter how many fish I caught or how big they might be; it mattered that they were thriving here and now—and not just surviving.

Cari rowed upriver to meet me, and we stood together in the river with the water passing around our knees and the clouds floating peacefully over our heads in a bright blue sky. We sipped our drinks and ate our sandwiches and never felt the need to speak—just be. But the sun was slipping behind the eagle's nest bluff and the ocotillo were glowing those final warming sunbeams. The temperature began to drop and Cari started to shiver, so we loaded into the raft and crossed back over toward the still sunny side of the river, away from the high limestone cliffs and toward the sycamore trees adorned in autumn gold. I still had one important action to complete, and it had nothing to do with catching a Devils River smallmouth. I had promised my dear friend Emily Whitlock that I would release some of her loving husband Dave Whitlock's ashes into a river that was dear to his heart—the Devils River.

As we crossed over into the warming light, Cari asked, "Did you decide where you want to release Dave's ashes?" Being the Imperfect Texan Buddha and Warrior-Poet that I am, I was looking for a sign from the universe, and I received it. When I looked up into the last of the waning daylight, I saw the first bald eagle that I have ever seen in all my years in Texas. Like that first beaver on the Llano River, I have seen bald eagles from Alaska to New York but never in my beloved Texas Hill Country. And this was without a doubt the most majestic eagle I've ever seen, with an angelic wingspan and a prophetic presence as it landed on top of the cliff face and watched us with intensity and intention. This was the place for my friend Dave Whitlock to join the river once more. Perhaps this place in the river should be called "Dave's Crossing."

I stepped out of the raft and onto a slightly submerged limestone island where the currents moved quickly around each side and onward toward the Rio Bravo. When I opened the bottle and freed Dave's ashes into that current, I smiled and said, "Lead the way, my friend. I will join you here whenever my time comes to break the tippet and cross over." After all, "Raindrops need not fear the river. The river is our Home."

We were high up on the desert mesa when my friend Cari Ray and I stepped out of our trucks and into an ancient world of flickering fire-light and mystical memories. I had arranged for us to be guided to the "Sunburst Shaman's Cave" by a woman who has invested much of her life to connecting people to both the past and present of this land and its spiritual soul. She was an older woman of indigenous Mexican American descent, who dressed herself in earth-toned clothing and wrapped her braided black and gray hair with some sort of naturally sourced twine. She did not smile, and her demeanor was cordial but aloof. I had the sense that she was staring into and through me as if she was trying to decide who or what I was and how to deal with me. As we walked she told us stories of her grandmother being what might in northern Mexico be called a *curandera*, or healer.

Cari later commented that walking with her felt like "walking in and out of clouds . . . or in the presence of a lantern swinging in the wind, giving intermittent darkness and light." I sensed in her a desire to be seen as a healer and person of light. I was struck by the feeling that she too needed to be healed. (Don't we all?) So many times in life our teachers come in unlikely forms. Sometimes a stone is the floor and sometimes it's the ceiling.

So, I followed her over the mesa's brow and down its rocky shoulders into the canyon where the ancients once walked and the future seems uncertain. As I walked the narrow trail I could imagine the first people of this canyon walking over these same stones just in front of me. After a while their image vanished, as if they had gone ahead, and I found myself alone with my thoughts and the serenity of birdsong and breezes. I felt as if they were waiting for me or, perhaps, I was waiting for them to remind me of something vital that I had forgotten. I am not immune to the amnesia that comes with society's illusions and miss-stories.

The first time I camped along the high desert mesas of the Devils River, I was seeking solace and guidance after the ending of the world I thought I knew and the beginning of the world I was coming to know. I felt then as if my soul had been kicked out of me, and it seemed natural to invest some time wandering around, looking up and down, and trying to find exactly where I'd lost it and how to put it back. Even then I knew that we are souls with bodies, not the other way around, and without the essence of who we are—intact and authentically within our vessels—we are unblinking, inanimate objects; we are no longer soil, we are dirt. But since then, the universe has taught me many things that I most likely already knew at birth but had forgotten with the acquisition of ego and the loss of oneness. Now I see that I had never lost my soul, only the illusions I once trusted as fact, and the fictions I once accepted through faith.

As we walked, she told us of her upbringing on both sides of the imaginary line we call "the border." She spoke of growing up in Del Rio, Texas, and of living a few months each year in Tamaulipas, Mexico, where her family had a sheep farm. She told us of her grandmother and mother passing down to her the culinary, textile, and medicinal knowledge of the ancients. She learned how to make dye and soap from the roots of plants; food from their flowers, fruits, and flesh; and medicine from every imaginable part of the plants. When she touched the leaves of the plants, I imagined my grandmother's hands holding her rosaries. It was less about holding and more about embracing; it was more of a loving, comforting caressing of another living being. As she described each plant and each feature of the living landscape, I saw perhaps the one thing our guide was able to genuinely love and find comfort in—this place, these living beings, and the imagined lives of "the ancient ones." Where foreign eyes saw only desert and death, indigenous people saw a Garden of Eden. They saw Home.

They were right to see this land as plentiful with gifts to offer. Along this southern stretch of the Devils River, four ecoregions mingle: Edwards Plateau, Trans Pecos, South Texas Brush Country, and a smattering of Tamaulipan thorn forest. We walked among the bright yellow splatters of parralena flowers and the lavender splendor of Cenizo. The ocotillo was covered in bright green leaves, while the sotal, lechuguilla, and Spanish dagger plants stood silent in the high desert wind. Prickly pear, tasajillo, and hedgehog cactus grew from the dry top of the mesa to the rocky ridges and walls below, where argarita showed off its sweet red fruit and sharp yellow spines. It is said that in Texas everything outdoors will "stick, stab, sting, or bite you" and I've found that to be quite true. Or as our indigenous guide remarked, "This is a tough land and if you live here, you become tough too."

Halfway down Sunburst Canyon's wall, we stopped to drink some water and take in the scenery. I allowed my eyes to follow the contours of the landscape from where we stood all the way down to the Devils River. I could see the high mountains of Mexico that seemed close enough to touch, although they might as well have been on some distant unreachable planet. We humans have a penchant for creating lines in the sand. We create our own heavens and hells.

As we began to walk farther down toward the canyon floor and its creek that intermittently carries life-giving water to the Devils River and on to the Rio Grande, I began to remember the paradise that Cari and I have found here in this timeless, magical, delicately brutal place. As I thought of this sacred place and its faraway people, I felt long-ago feelings that I could not comprehend. We walked out into a clearing in the forest and all at once found ourselves standing in front of the Sunburst Shaman's Cave.

Across its limestone face were the paintings of shining suns and singing shamans. There were symbols and talismans whose meaning may be lost to the ages, and yet, like hearing a song in another language, the melody spoke to me just the same. Across the mouth of the cave was a natural garden of native plants, and our guide pointed to each one while describing the many gifts they offered. Edible flowers, fruits, flesh, and roots. Medicinal teas and tinctures that were believed to alleviate everything from asthma to arrhythmia. And leaves that became sleeping mats and branches that became digging tools and roots that became dyes for sacred symbols painted on the cave walls, symbols that tell the story of long-ago gatherings where communion with the outer world was sought and community with the inner world was found.

Along the walls of the cave were various versions of sunburst drawings that clearly looked like a child's drawing of the sun—a circle with radiating beams of light. It struck me how universal art can be when it is simple and direct, without pretentious embellishments or subjective nuance. Many of the paintings were of shamans and otherworldly figures or of symbols whose meanings have been lost with the passage of time and tribes. But the sunbursts were universal and timeless, and my eyes gravitated toward each one.

While our guide described the illustrations, I noticed that she repeatedly referred to the shaman figures as being female. I found that interesting, and I wondered if she knew something about, or simply imposed something upon, these ancient figures. But the one thing that stood out most was her repeated reference to the existence of "community" and her insistence that because of their tight community these people not only survived here, they thrived here.

At one point we laid down on our backs looking up at the cave ceiling as its former residents might have done some four thousand years ago. On the roof were more paintings of sunbursts and shamans and the many symbols of long-ago meaning. And then we walked along the cave's length to where it transformed from the place for the ceremonies of the dead to the space for the community of the living. Here we could see how the stones of the cave's floor had been tossed by hands of several thousand years past, so that people might sleep on softer ground and dig earthen ovens to bake their sotal root dinners. At the end of the living space was a tinaja—a pocket of surface water carved out of the stone by centuries of rainwater erosion. It seems that the ancient ones had everything they needed to thrive—food, water, shelter, medicine, and a deep sense of a living community. I was struck by the impression that in these current times we take the first four needs for granted and are losing the final one to the emptiness of "the grind."

Just before we left the Sunburst Shaman's Cave, our guide asked me the only question she ever asked; up to that time she had been consumed in telling us what we "should be seeing" and "should know." It felt like Sunday school where the teacher is simply repeating all the things they were told and accepted without question or reflection. But now as we looked back at the entirety of the cave and its paintings, she asked me, "Why do you write?" I paused for a moment before gesturing toward the images of ancient sunshine and long-ago lessons. Pointing to the cave I said, "This is why I write. I write for the same reason that they drew and painted. They are speaking to me now from the walls of their long-ago home. I write because the symbols that we call words will outlive me and may help guide other living beings in their journey, just when they need it most. I write because it's all I have to give, and without giving, I'm not really living."

The first time I visited a Devils River rock art site was with my daughter, long ago. The park ranger who took us there did not tell us anything; he simply allowed us to be in the space and experience the spirit of the place and the people who once lived there. It was a sacred moment in my life, and I appreciated his sensitivity. He had no agenda but to guide us to the cave, not through the cave. But every experience has its unique lessons, and my life's teachers have come in many forms. I am grateful for the hardness of stone and the softness of water. Both have their own forms of power and grace.

Then as now, words are my petroglyphs. I scratch them out on a more or less durable rock. I paint them in colors wrung from Nature like squeezing out my dreams and fingertip painting on my limestone-self. They reflect and sustain my most noble nature. They are my tapping on the prison wall.

There is a Spanish word that I first encountered while fishing in New Mexico, and it seems to be used here, too, and by the people of northern

Mexico. The word is *querencia*. It doesn't seem to have an exact translation into English but is used to mean the place where our strength is drawn from, where we feel most at home and are able to be our most authentic selves. I wish the English language had deeper forms of so many words. Words like "love," and "living" and "Home."

It seems to me that we have lost our way. We can never get home by clicking our heels or gathering facts—we must authentically feel it to find it. If we treat the earth like "dirt" and the rivers like a "resource" or each other like "the other," we remove all chance for empathy, understanding, and mutual respect. The death of empathy and our willful separation from the natural world is leading us to our collective doom and taking the world with us. Every living being has an immune system, and Nature does too. With a changing climate and the ensuing increased threat to humanity and all living things, Mother Nature may be fighting back now. The earth is a living being, as is each river, fish, and microbe. There is a consciousness in the eyes of the fish in my hands. She wants to go home.

My friend Cari once said, "There are those we walk by . . . and those with whom we walk." Over my lifetime I have done my best to find and build my tribe, my community, my hopeful world of imperfect but empathetic and humbly bold souls with whom I can walk. Nature and the act of fly fishing has been my gateway to this quest. Standing in a river I feel connected to the universe. I don't count fish by number, weight, or length. I don't value the experience solely by the fishing. I practice being present, alive, aware, and grateful with every cast and every current. And in the end, this story has no ending.

As I stood in the clean, clear waters of the Devils River and watched the eagle flying overhead while freeing the ashes of my friend into the currents and eddies, I was struck by the feeling that I was home. It's been a long journey that consisted of more than floating and fishing from Alaska to New York. It's been an adventure in discovery about the meaning of health and happiness and what it means to be Home. And I discovered that Home really is wherever my broken heart beats most truly. It is where my family and friends are. It is where my juniper and oak canyons and spring-fed rivers embrace me—my beloved Texas Hills. The earth is my Home. Nature and the best of human nature is where I belong. Empathy and understanding are the bricks and mortar that bind our house and protect us from our lesser selves. I discovered that the one most important thing we must all realize and remember is that there is no "other." We Travel Together, y'all.

Epilogue

If you do not change direction, you may end up where you are heading.

—Lao Tzu

There are times when the waters and currents of our lives are troubled and tremble with the fear and sense of hopelessness that comes when there are wars and rumors of war, pestilence and plague, rivers and oceans turning acid, glaciers and polar ice caps melting, sea levels rising, storms and firestorms becoming more frequent and intense, and forests, fish, songbirds, butterflies, and so many other forms of wildlife vanishing into oblivion, and we the people of the earth seemingly unable or unwilling to turn away from those fictions we've created to justify the darkest recesses to our beings. These are such times. None of which I just wrote is fiction—it is the current reality we are creating. Yet somehow, I hold on to the glimmer of hope that together enough of us will build a better tribe—one that will illuminate a path of redemption and in doing so negate the one we now follow.

The earth is our one true Home. But home is also that place in our hearts where love, comfort, and community come together. Home is a place and condition of our own choosing and creation. Paradise is found and lost on a whim. We can do nothing about what happens to us in life, but we can choose how we will respond (not react) to our conditions. Almost everything is within our power, and absolutely everything is survivable, except the last thing.

I choose to remain hopeful that the people who truly love this Home will come together as one tribe and discover new and better ways to live as part of Nature—instead of pursuing the self-destructive path of separation from Nature. Together we can choose to be the ocean and not the fish. One tribe, one Home. We are One.

Acknowledgments

One of the great joys of this adventure has been sharing it with some of the best people on earth. I am grateful to each of my friends who appear within these pages as well as my behind-the-scenes friends—Janice "Lil Red" Bowden Hardaway, Maggie Serva, Catherine Homberg Gerch Aileen Hitomi Lane, Tiffany Dawn Compton, and Cari Ray—for their inspiration, support, and kindness. Without each of these wonderful people, this work, and my life, would be so much the lesser.

I am deeply grateful to my editor and friend, Gene Brissie of Lyons Press, for taking the time to read my story, and then for choosing to bring it from the original manuscript to the completed literary work you now hold. And I appreciate the guidance and support of my production editor Meredith Dias and my copy editor Ann Seifert, as well as Max Phelps and Alyssa Griffin at Lyons Press. They have made this long journey so much more pleasurable and meaningful with their professionalism, knowledge, and heartfelt dedication.

Most of all, I am grateful to my wife and best friend of forty years, Alice, our amazing daughter Megan, and her wonderful partner Nick, who together form our tight little family that we refer to as "the Regiment." And although I love my entire Regiment, I would be remiss not to single out my wife Alice, who has proofread and given constructive input to every word of everything I have ever written and published. With a lifetime of dear family, friends, and fishing, I am a fortunate man indeed.